# Elements of Aerodynamics

A Concise Introduction to Physical Concepts

*Oscar Biblarz*
*Naval Postgraduate School*
*Monterey, California, USA*

This edition first published 2023


*Registered Office*
John Wiley & Sons, Inc., 111 River Street, Hoboken, NJ 07030, USA

For details of our global editorial offices, customer services, and more information about Wiley products visit us at www.wiley.com.

Wiley also publishes its books in a variety of electronic formats and by print-on-demand. Some content that appears in standard print versions of this book may not be available in other formats.



*Library of Congress Cataloging-in-Publication Data applied for:*
Hardback ISBN: 9781119779971

Cover Design: Wiley
Cover Images: © Volodymyr Burdiak/Shutterstock; Gergitek Gergi tavan/Shutterstock; Courtesy of Oscar Biblarz

Set in 9.5/12.5pt STIXTwoText by Straive, Pondicherry, India

**Elements of Aerodynamics**

# Contents

# To the Student

Aerodynamics is considered to be part of fluid mechanics and as background some college level exposure to basic physics and chemistry on your part will be necessary. Additionally, a background in calculus and basic thermodynamics will be assumed. Specifically you are expected to know the following:

1) Differentiation and integration, logarithms, exponents, and trigonometric functions
2) The meaning and symbols for total and partial derivatives
3) Vector calculus, the meaning of dot and cross products
4) How to draw and interpret free body diagrams
5) How to resolve vectors into components in both Cartesian and polar coordinates
6) Newton's second law of motion and its proper units as related to force and mass
7) Properties of gases such as density, pressure, and viscosity
8) The laws of thermodynamics

The first five prerequisites are mostly math; the last three cover more physical principles. A background in fluids and in thermodynamics is important for our studies and a course in *gas dynamics*, either prior or concurrent, is very helpful. If you have not recently had such courses, you may want to pay special attention to the first four chapters of this book along with Chapter 8. Equations introduced there will be used in the rest of the book. Cartesian coordinates are our default notation where $x$ is along the relative free stream flow direction and $y$ can be directed either vertical in two dimensions or spanwise in three dimensions, and the distinction must be kept clear in context. You should get some confidence on the subject matter from studying the examples and by working problems at the end of each chapter. Chapter 2 considers why we prefer to use non-dimensional quantities in aerodynamics and is strongly recommended reading. Throughout the book, you will find some emphasis on interpretation over derivation and that many chapter examples and some problems contain proofs of the topics in which they are embedded.

Chapter 1 includes an appended *Glossary of Terms and Symbols* used throughout the book, notation that forms the backbone of aerodynamics analysis, and these terms appear frequently in all chapters. In Chapters 2 and 3, we begin with the fundamental laws in their *control volume form*. If you have had a proper course in fluid mechanics, much of this material should be familiar to you. Constant-density fluid relations are examined in Chapters 3 through 7 because much of aerodynamics is incompressible and because of the important theorems that relate to such incompressible flows. In Chapter 8, you are introduced to the background and characteristics nature of compressible fluids. Because several special concepts are developed in Chapters 8 and 9 that are not treated in many thermodynamics and fluid mechanics courses, *read these chapters with care even if you have the relevant background.*

Chapters 5 and 6 are dedicated to the incompressible aerodynamics of thin airfoils and cover most of the important results in this field. Chapter 7 succinctly discusses viscous effects and in particular the boundary layer concept since it applies to all flow regimes; the equations describing viscous effects are nonlinear, and dedicated software is often needed to go beyond our presentation in this chapter. Chapters 8 and 9 deal with compressible subsonic and with supersonic domains. Chapter 10 wraps up our identified flight regimes by discussing transonic and hypersonic flows; these require separate treatment because of their complexity.

Chapter 11 returns to incompressible flows to include details and applications of high lift airfoils. This chapter has been separated from Chapters 5 and 6 because it approaches the subject in a less conventional manner and because in it we comment and evaluate the present state of the field thereby underscoring some theoretical possibilities in aerodynamics.

Appendix A shows an abbreviated form of the Standard Atmosphere table. Appendix B presents the isentropic and oblique shock contents of our *Aerodynamics Calculator* that is companion to this book and uses the same notation. Also, some free software programs available from the Internet are mentioned in Appendix B since this resource has a lot of growth potential. Appendix C complements Chapters 5 and 6 by including traditional detail for the Fourier series approach. *Selected References* are organized under the following general topics: Aerodynamics, Fluid Mechanics, Gas Dynamics, and Thermodynamics. We include answers to selected problems before the Index.

This book has been written as an entry-level text and Chapter 1 details much of the foundation of this subject. Once you have passed the first chapter, the remaining chapters follow a similar format. When you start each chapter, read the introduction, as this will give you the general idea of what the chapter is all about. The next section contains a set of learning *Objectives* (beginning with Chapter 2). You should aim to comprehend the listed subjects after completing the chapter. A few objectives are marked *optional* as they are only for the most earnest students. Merely scan the objectives, as they will not mean much at first. However, *they will indicate important things to look for.* As you read the text material, you may occasionally be asked to do something – complete a derivation, fill in a chart, draw a diagram, etc. Active participation will help solidify important concepts and provide you feedback on your progress. At the end of each chapter, you will find sections on Problems and Check Test – the latter is meant to complement the Objectives section. Starting from Chapter 4, we finish off each chapter with a section called *Enrichment Topics* – these are meant to enhance the information in the chapter in non-traditional ways.

The author solicits input from students and other users, particularly suggestions meant to improve the contents of this book. The scientific principles of aerodynamics have been around for many years and its hydrodynamics and mathematical legacies are classic and have remained pretty much unchanged (except for the use of computers). The nature of textbooks has evolved with the digital age, and it is hoped that the companion website and easy access to the internet will be helpful to the reader.

# Preface

*Source:* Ensign John Gay / Wikimedia commons / Public domain.

This is a 151 F/A-18 *Hornet* aircraft generating shock waves upon breaking the sound barrier.

This book attempts to blend the purely classical mathematical foundations of aerodynamics together with real-life flight details in a coherent package without dwelling into details of purely academic interest. While other books introduce this subject from a general fluid mechanics point of view, my preference has been to first present what makes aerodynamics unique. The role of viscosity needs to be identified early because, aside of producing drag, viscous effects properly focused at the airfoil's trailing edge are an essential ingredient for lift generation (the classical Kutta condition in subsonic flows). Flow circulation plays an indispensable role in both lift and some of the drag experienced by airfoils and vorticity cannot be generated in ideal fluids, so focused circulation conditions need to be treated as the key ingredient in aerodynamics. These effects manifest themselves in a favorable way for a flight-restricted operating region mostly at low angles of attack. Moreover, the extensive use of dimensionless coefficients in aerodynamics requires discussion of not only how they are obtained but also how they may be properly used. Coefficient definitions must be carefully scrutinized when analyzing lift, drag, and moments dependencies because the standard usage of incompressible definitions often contains the very variables we are examining. Also, because of the prevalence of flow separation, *nonlinear thin-airfoil theory* has heretofore not been considered a viable subject for an introductory textbook but, as I aim to show, this somewhat narrows the potential of the subject.

Aerodynamics deals with forces together with the moments they generate on objects in flight. Of these, the aerodynamic lift force is of particular importance in that lift is essential for nearly all vehicles moving in air in contrast to those moving in water where most objects can remain buoyant. Lift has a uniquely desirable mode of operation where viscous regions thinly drape around lift-producing wings so that the bulk of the flow remains ideal and amenable to closed-form theoretical descriptions. My aim has been to feature the study of slender streamlined objects in steady flow where viscous effects can be relegated to thin boundary layers. Unless stated otherwise, conditions under which the boundary layers remain attached are always assumed. Air behaves as an incompressible fluid at low enough speeds and in practice all flying objects must operate in this region either partly or fully, so this flight regime is always important. Properly understanding incompressible flow behavior also helps to develop some of the features of compressible flow. Chapter 11 extends standard incompressible flow methodologies and has been located at the end because its approach is less conventional than the preceding 10 chapters.

This book is the result of many years of teaching basic aerodynamics at the senior-undergraduate and first year graduate level and includes some of my transonic-flow research activity. The classic gas dynamics books by A. H. Shapiro and by H. W. Liepmann and A. Rosko as well as more recent books on aerodynamics by A. M. Kuethe and C. Y. Chow, J. J. Bertin, and J. D. Anderson have influenced my approach on this subject. I have adopted the same topic organizational technique found in *Fundamentals of Gas Dynamics* which I co-authored with R. D. Zucker, and which is based on *educational technology* principles; my subject development differs somewhat in that certain proofs are given as examples and I have added enrichment topics. Mr. D. F. Dausen at the Naval Postgraduate School made many helpful suggestions. Mr. A. I. Biblarz developed the *Aerodynamics Calculator* as a direct companion to this book; details of this calculator are given in Appendix B.

Oscar Biblarz
Monterey, CA, USA

# About the Companion Website

A companion website with additional resources is available at:

**www.wiley.com/go/elementsofaerodynamics**

The materials in the companion website includes solutions manual and review questions for each chapter.

# 1

# Introduction and Approach

## 1.1 Introduction

### 1.1.1 Wing Aerodynamics

Airplanes feature many external components, but by far the most important ones are the *wings* that provide the *lift* necessary to remain airborne. Wings also produce *pitching moments*, and much of the drag on the airplane, therefore, this book focuses on *the aerodynamics of the wing*. In our approach, we examine flows that surround the airfoil with thin viscous layers that create minimum drag as it generates lift. Under fully immersed conditions, moving objects may be studied with irrotational flow descriptions (e.g. Bernoulli's equation) in relatively conventional ways. As will also be evident, the lift along with portions of the drag and other key aerodynamic phenomena can be analyzed with a flow-induced vorticity model known as *circulation*, and this important concept is properly introduced in Chapter 3 and applied to airfoils beginning with Chapter 5.

Our most frequent aerodynamic situation focuses on airfoils restricted to operating angles of attack where no flow separation develops. Airfoils are assumed to be thin and either symmetric or of low curvature in subsonic flows so that we may replace them by their *mean line* because effects of thickness are known to play a secondary role. Low-viscosity air flows develop mostly "thin shearing boundary layers," and while these remain attached to the shape of the body closely resembles the displaced inviscid region surrounding it. Viscosity, however, plays an indispensable role in generating the above-mentioned *circulation* that surrounds lifting airfoils because it depends on a stagnation point developing at the airfoil's sharp trailing edge (the so-called the Kutta condition). Airfoils cannot produce much lift whenever their boundary layers detach and lift does not occur on moving immersed objects without any embedded circulation. Flows of interest in aerodynamics, however, while never really *inviscid* may be tailored so as to isolate the effects of viscosity while allowing the generation of lift through purposely streamlined airfoil configurations.

The background necessary to study aerodynamics assumes you have had previous formal introduction to fluid mechanics preferably followed by gas dynamics, together with the needed exposure to basic thermodynamics, but an effort will be made to briefly review such concepts as they are presented. Courses on fundamental calculus and modern physics are needed prerequisites. Our listings under Selected References contain many of the textbooks that cover these subjects. At the end of *this* chapter, we present a **Glossary of Terms and Symbols** which are commonly used in aerodynamics intended to be a useful and handy reference for the rest of the book.

*Elements of Aerodynamics: A Concise Introduction to Physical Concepts*, First Edition. Oscar Biblarz.

Companion website: www.wiley.com/go/elementsofaerodynamics

## 1.2 Necessary Assumptions

While we shall endeavor to develop wing-aerodynamics with some generality, for most situations we restrict ourselves to one or more of the following conditions:

A) Steady flows – effects of mass inertia due to transients are neglected as we concentrate on wings traveling with a time-independent velocity so aircraft are moving under non-maneuvering situations.
B) One-dimensional or two-dimensional flows – flow descriptions primarily based on $(x, y)$ or $(r, \theta)$ coordinates.
C) Negligible body forces – the weight of the air mass surrounding the airfoil is reflected in pressure and density variations with elevation in the Standard Atmosphere table.
D) Incompressible flows – here changes of density can be neglected in the fluid equations; formulations resemble those in hydrodynamic flows (i.e. objects immersed in water).
E) Compressible flows – here effects of compressibility become important and are separated into subsonic, transonic, supersonic, and hypersonic regimes depending on Mach number.
F) Inviscid or ideal flows – these flows operate where shearing forces may be neglected (outside of a region labelled *the boundary layers*).
G) Boundary layer flows – viscous effects are localized in a thin envelope around the airfoil where the flow is largely incompressible even at high Mach numbers.
H) Streamlined flows – airfoils with properly *rounded leading edges* and *sharp trailing edges* in subsonic flows and pointed leading and trailing edges in supersonic flows.

Briefly then, our *default approach* will be the study of thin, streamlined airfoils operating with attached boundary layers. Whenever we extrapolate our discussion to an entire aircraft, we will consider a cruising or force equilibrium situation where lift equals weight and thrust equals drag as shown with the vector lengths in Figure 1.1.

## 1.3 Units

Aerodynamics uses concepts from both mechanics and thermodynamics so it is necessary to be proficient with two systems of units, namely, the metric or International System (SI) and the English Engineering (EE) system. You will need to manage gas properties in both systems, even if we work mostly with the SI system. In aerodynamics, we develop many important expressions with dimensionless coefficients because they are devoid of units and more general than their dimensional counterparts.

Gas properties such as pressure, density, and the speed of sound, consist of several *basic units*, namely, length, time, mass, and temperature. These units are listed in Table 1.1. Lengths exist in up to three dimensions as vectors, whereas the rest are scalars.

### 1.3.1 Additional Systems of Units with Their Conversion Factors

We also need to be aware of Gravitational Units such as the "lbm" and the "kgf" together with their equivalent mechanical units the "slug" and the "newton." Thermodynamics often uses units not generic to fluid mechanics, called "caloric," which require conversion factors to SI or EE units. Another factor is $g_c$ as it appears in Newton's second law (i.e. $F = ma/g_c$). In the gravitational

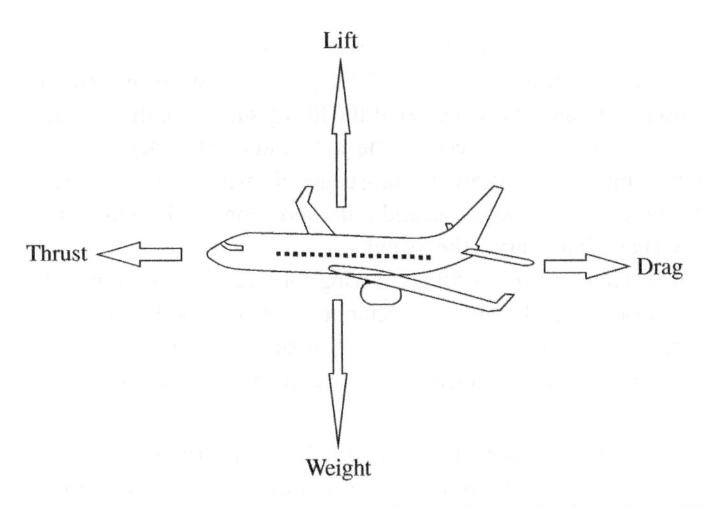


**Figure 1.1** Generic commercial aircraft on cruise mode with principal forces acting on it.

**Table 1.1** Basic systems of units.

| Dimension | International System (SI) | English Engineering (EE) |
|---|---|---|
| Length | meter (m), centimeter (cm) | foot (ft), inch (in) |
| Time | second (s) | second (sec) |
| Mass | kilogram (kg) | pound-mass (lbm), (slug) |
| Temperature | Absolute: kelvin (K),<br>Relative: Celsius (°C) | Absolute: Rankine (°R),<br>Relative: Fahrenheit (°F) |

system, mass and weight amounts are equivalent (often undistinguished) because the numerical value of the acceleration due to gravity at sea level (i.e. the $g$ in $W = mg/g_c$) cancels out.

$$\text{Gravitational } g_c \quad g_c = 32.174\,\text{lbm-ft/lbf-sec}^2 \quad \text{or} \quad 9.807\,\text{kg-m/kgf-sec}^2 \tag{1.1}$$

$$\text{Mechanical } g_c \quad g_c = 1\,\text{slug-ft/lbf-sec}^2 \quad \text{or} \quad 1\,\text{kg-m/N-sec}^2 \tag{1.2}$$

Another conversion factor needed when dealing with the units of work and heat with the first law of thermodynamics is "*J*," the mechanical equivalent of heat.

$$J = 778\,\text{ft-lbf/Btu} \quad \text{or} \quad 4186\,\text{joule/kcal} \tag{1.3}$$

We should also note here that among thermodynamic units the "joule or J" is defined as a N-m (or 1.0 kg-m$^2$/s$^2$) in the SI system, whereas the "Btu" is defined as 778.16 ft-lbf in the EE system. While thermodynamic work and heat transfer often share the same energy units, they carry different entropy contents.

## Notation

A considerable number of special terms are needed to describe and analyze flows around *airfoils* (see the **Glossary**). Here we mention a few wing related ones:

The cruising or *x-direction* – the direction of the *relative wind velocity vector* ($\overrightarrow{V_\infty}$).

Wingspan ($b$) – the span or width of the wing in the *y-direction*. This span is the distance between wing tips without the aircraft's fuselage so as to only represent the lifting surface of the aircraft. (Note under two-dimensional conditions we use the coordinate $y$ instead of $z$ for elevation.)

Wing chord ($c$) – this is the line connecting the leading and trailing edges of a wing. For trapezoidal and other non-rectangular wings, there can be a *root chord* and a *tip chord* along with a *taper ratio* ($\lambda$). The chord dimension and the span characterize the airfoil.

Planform area ($S$) – the total wing area. There are many different wing configurations that include rectangular, elliptical, tapered, swept-back, and delta. For rectangular planforms, $S = bc$.

Aspect ratio ($\boldsymbol{AR}$) – defined as the ratio $b^2/S$; for a rectangular wing, it simplifies to $b/c$.

Angle of attack ($\alpha$) – represents the orientation of the wing planform area to the relative wind velocity vector ($\overrightarrow{V_\infty}$).

Mean camber line – a line halfway between the upper and lower surfaces of the airfoil. When this line is not straight but "concave down," the airfoil is said to be *cambered*, a shape which has beneficial effects for subsonic flight.

Sweep angle ($\Lambda$) – the angular orientation of the planform's leading edge with respect to the coordinate perpendicular to the relative wind direction ($y$) and is usually given the symbol $\Lambda$. It delays effects of compressibility that affect maximum lift and stall.

Thickness and thickness distribution – variables mainly affecting the viscous boundary layers. Maximum airfoil thickness ratio is usually written as $t_m/c$ and which seldom exceeds 12%.

## 1.4 Equation of State and Fluid Properties

### 1.4.1 Perfect Gas Equation of State

The perfect gas equation of state accurately represents atmospheric air except at elevations near the "transatmosphere" or "boundary with space." This equation says that the ratio of the absolute pressure to the product of the density times the absolute temperature is a constant that depends only on the molecular constituents of air (which are mostly nitrogen and oxygen).

$$\boxed{p = \rho RT} \tag{1.4}$$

| | **(SI units)** | **(EE units)** |
|---|---|---|
| $p$ = absolute pressure | N/m$^2$ or Pa or bar<br>megapascal (MPa) = 10 bar = $10^6$ N/m$^2$<br>1 atm = 1.013 bar = 760 mm Hg | lbf/ft$^2$ or psia (lbf/in$^2$),<br>1 atm = 14.7 psi = 29.9 in of mercury |
| $\rho$ = density | kg/m$^3$ | lbm/ft$^3$ or slug/ft$^3$ |
| $T$ = absolute temperature | K<br>$T$(K) = $T$(°C) + 273.15 | °R<br>$T$(°R) = $T$(°F) + 459.67 |
| $R$ = gas constant | 8314/M.M.<br>287 N-m/kg-K (for air) | 1545/M.M.<br>53.3 ft-lbf/lbm-°R (for air) |

*Note:* The molecular mass (M.M.) depends on the gaseous constituents as given in the Periodic Table of the Elements and has the same numerical value in both systems of units (e.g. for air, M.M. = 28.97 kg/kg-mole or lbm/lb-mole). The "universal perfect gas constant" is 8314 in SI units and 1545 in EE units.

**Example 1.1** Calculate the gas density inside a toy balloon filled with helium at room conditions.

We assume that the helium in the balloon is at a temperature of 20 °C (293.15 K) and at atmospheric pressure ($1.013 \times 10^5$ N/m$^2$). In order to calculate helium's gas constant, we need to divide the universal gas constant by the molecular mass of helium, i.e. 8314/4.0 = 2078.5 N-m/kg-K. Hence the density becomes

$$\rho = \frac{1.013 \times 10^5}{2078.5 \times 293.15} = 0.166\ \text{kg/m}^3$$

The density of air under these conditions is 1.22 kg/m$^3$ and such density difference leads to a "buoyancy force" so the helium balloon would tend to rise. Note that the actual pressure inside the balloon should be slightly higher than the ambient pressure.

Both pressure and temperature can be either relative or absolute, but unlike temperature the same units are used for pressure (with an "a" often added for absolute values in the EE system, i.e. psia or psfa). When the pressure is relative to ambient pressure, it is called *gage pressure.* But in most formulations, such as the equation of state, we must use absolute values of pressure (based on a vacuum condition, see Figure 1.2) and absolute temperature (based on absolute zero). Standard atmospheric pressure refers to a value at sea level commonly written as 1 atm.

The speed of sound is particularly important in compressible flows. It represents the speed of very weak pressure waves as they propagate through any medium without changing any of its properties. Denoted by the symbol $a$, for air as a perfect gas, it can be shown to be

$$\boxed{a = \sqrt{\gamma R T}} \tag{1.5}$$

$$\text{In air,}\quad a \approx 20\sqrt{T(\text{K})}\ \text{m/s} \quad \text{or} \quad a \approx 49\sqrt{T(°\text{R})}\ \text{ft/sec}$$

The gas constant ($R$) comes from Eq. (1.4). In the EE system, we need to multiply the value of $R$ by the factor $g_c = 32.2$ lbm-ft/lbf-sec$^2$ to arrive at the proper units. The value of the ratio of specific heats for standard atmospheric air is $\gamma = 1.4$.

**Example 1.2** Calculate the speed of sound in air at 70 °F using standard values for $\gamma$ and $R$.

We will use Eq. (1.5) in the EE system to illustrate the need to change units used in thermodynamics to mechanical units. For air, $R = 53.3$ ft-lbf/lbm-°R and with $\gamma = 1.4$, and we find the absolute temperature from $T = 459.67 + 70 = 529.67$ °R.

$$a = \left[1.4 \times 53.3(\text{ft-lbf/lbm°R}) \times 32.174\left(\text{lbm-ft/lbf-sec}^2\right) \times 529.67\,°\text{R})\right] = 1{,}128\ \text{ft/sec}$$

This calculated magnitude of 1128 ft/sec or its equivalent 343.7 m/s is a useful number to remember.

### 1.4.2 Viscosity

All fluids (gases and liquids) continually deform under the action of tangential or shear forces in contrast to solids that begin to deform only under extreme force magnitudes. Viscosity is a fluid

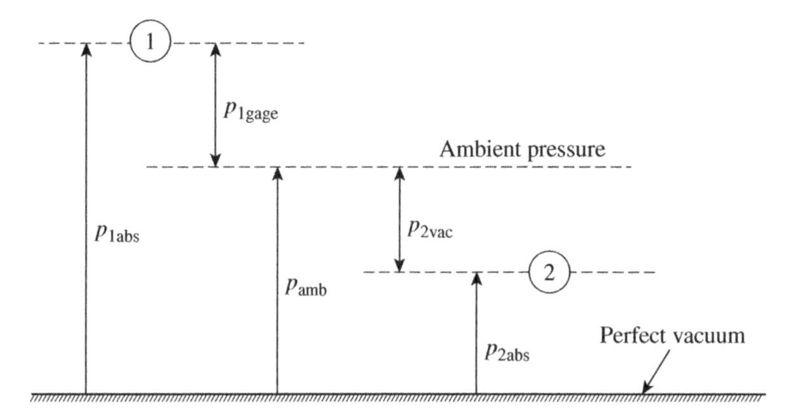


**Figure 1.2** Absolute, gage, and vacuum pressures.

property through which a shearing stress ($\tau$) is transferred to an immersed object's surface under the action of a velocity gradient ($du/dy$). Denoted by the symbol $\mu$, viscosity is defined as

$$\mu \equiv \frac{\text{Shear stress}}{\text{Rate of angular deformation}} = \frac{\tau}{du/dy} \tag{1.6}$$

In air, $\mu = 1.8 \times 10^{-5}\ \text{N} \cdot \text{s/m}^2$ or $3.8 \times 10^{-7}\ \text{lbf} \cdot \text{sec/ft}^2$

The numerical values shown here are for air at standard temperatures. Viscosity is independent of pressures, except in the hypersonic regime, but it increases in gases as the temperature increases. Note that the amount of deformation is of less significance than *the rate of deformation*. The viscosity of atmospheric air is low relative to liquids and, since nearly all drag is due to shearing at the surface or skin friction, the role of good aerodynamic design is to minimize drag while maintaining a stagnation point at the rear of an airfoil (i.e. the Kutta condition). We discuss viscous effects in Chapter 7 in more detail.

In incompressible flows (i.e. those of constant density), we often encounter the ratio $\mu/\rho$ which is called the *kinematic viscosity* and its units simplify to $\text{m}^2/\text{s}$ or $\text{ft}^2/\text{sec}$. This ratio is given the symbol $\nu$ (see the Reynolds number in Chapters 2 and 7).

### 1.4.3 Perfect Gas Relations

A thermodynamic *process* represents two or more *states* within the medium. Several types of processes with perfect gases are of interest in aerodynamics and the "polytropic process", defined with a generalized exponent "$n$," is useful for writing perfect gas relations between properties as shown in Eqs. (1.7)–(1.9). For example, from Eq. (1.4) we can deduce that a constant temperature process leads to the ratio $p/\rho$ being constant so that $n = 1.0$ in Eq. (1.7). Other values of $n$ shown in Figure 1.3 require inputs from the laws of thermodynamics or are found empirically.

$$p\rho^{-n} = \text{constant} \tag{1.7}$$

$$\frac{T}{p^{(n-1)/n}} = \text{constant} \tag{1.8}$$

$$T\rho^{1-n} = \text{constant} \tag{1.9}$$

Equation (1.7) is sketched on the $p$–$v$ diagram shown in Figure 1.3 where instead of the density we show the specific volume ($v \equiv 1/\rho$) making Eq. (1.7) $pv^n$ = constant. Equations (1.8) and (1.9) may be implicitly found through the value of their "constant."

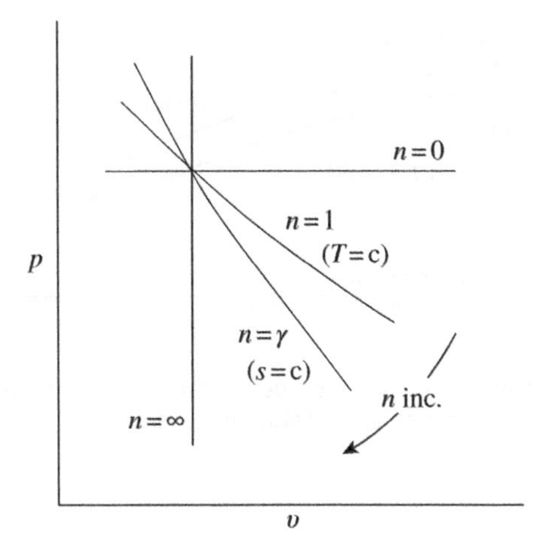


**Figure 1.3** Sketch of selected perfect gas processes. In the diagram, the notation "= c" means the property is "constant." Note that for $n$ = 0, $p$ = constant and for $n = \infty$, $\rho$ = constant.

### 1.4.4 Energy ($u$), Enthalpy ($h$), and the Ratio of Specific Heats ($\gamma$)

The *internal thermal energy* per unit mass of an ideal gas is given the symbol $u$ and the *enthalpy h* is related to it by an additional *flow-work term "p/ρ"* as given in the first law of thermodynamics. Moreover, in perfect gases these two properties are only functions of temperature via *specific heats* ($c_p$ and $c_v$). Numerical values given below are for air.

$$h \equiv u + p/\rho \quad \text{units : J/kg or Btu/lbm} \tag{1.10}$$

$$u = c_v T \quad \text{and} \quad h = c_p T$$

$$c_v = 716 \ \ \text{J/kg}\cdot\text{K} \ \text{ or } \ 0.17\,\text{Btu/lbm}\cdot{}^\circ\text{R}$$

$$c_p = 1000 \ \ \text{J/kg}\cdot\text{K} \ \text{ or } \ 0.24\,\text{Btu/lbm}\cdot{}^\circ\text{R}$$

The ratio of specific heats ($\gamma$) is defined in Eq. (1.11). In air at ordinary temperatures and pressures it remains relatively constant at $\gamma$ = 1.40 until gas temperatures begin to exceed 450 K (800 °R). Excepting hypersonic flows, we will use the 1.40 value for $\gamma$ exclusively in all our aerodynamic calculations.

$$\boxed{\gamma \equiv c_p/c_v} \tag{1.11}$$

### 1.4.5 Isentropic Flows

Entropy is a thermodynamics property introduced in the second law of thermodynamics. The specific entropy or entropy per unit mass is given the symbol "$s$" and has units of J/kgK or Btu/lbm°R. Because flows in most of aerodynamic analysis are adiabatic (i.e. no appreciable heat transfer takes place) entropy is seldom discussed. However, entropy is generated within shocks and inside the boundary layers by the flow's viscosity; as a guide for when we are excluding these two types of flows, we use the reversible-adiabatic or *isentropic flow* moniker. In isentropic flows, the thermodynamic equations for ideal gases very elegantly describe pressure, temperature, and density and will make routine use of them.

In Figure 1.3, the curve for $n = \gamma$ is particularly useful (recall $\gamma > 1.0$ for air) because it represents a constant entropy process (or $s = c$). Furthermore, along specific *streamlines* that form a stream tube (see Figure 1.4), a lossless or isentropic deceleration to zero speed coupled with a drop zero elevation leads to the useful reference thermodynamic state called the "total or stagnation state" which will be utilized in Chapter 8 and beyond. In this figure, $u$ is the internal thermal energy, *KE* the kinetic anergy, and *PE* the potential energy of the flow.

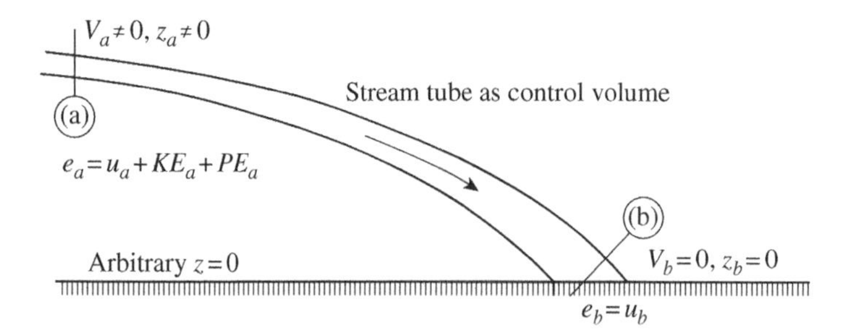


**Figure 1.4** Isentropic deceleration from location (a) to location (b) representing a process leading to the total or stagnation state. Total fluid energy ($e$) is conserved making $u_a + KE_a + PE_a = u_b$.

**Example 1.3** Air at 320 K and 1.5 bar undergoes an isentropic process during which its density is doubled. What are its final pressure and temperature?

For an isentropic process, $n = \gamma = 1.40$ and here we can use Eqs. (1.7) and (1.9) as follows:

$$p_2 = 1.5 \times (2)^{1.4} = 3.96 \text{ bar}$$

$$T_2 = 320 \times (2)^{0.4} = 422.24 \text{ K}$$

From the equation of state, Eq. (1.4), we find that $\rho_1 = 1.63 \text{ kg/m}^3$ so that $\rho_2 = 3.26 \text{ kg/m}^3$ demonstrating that an isentropic process lies between an isothermal and a constant density process as shown in Figure 1.3.

### 1.4.6 Mach Number (*Ma*)

Although the perfect gas equation of state (Eq. 1.4) applies during air movements, it gives no hint of flight conditions under which density changes can or cannot be significant. Aerodynamic regimes are best characterized in terms of the *Mach number* which is the ratio of the relative velocity ($V$) of an approaching flow to the local sound speed ($a$) at any flow location and is shown in Eq. (1.12); it is more formally introduced in Chapter 2. We will deal with the full range of Mach numbers throughout this book and more in depth in Chapters 8–10. Here we only need to distinguish important *flight regimes* in terms of Mach number because they help manage the way aerodynamic topics are presented (Table 1.2).

$$\boxed{Ma \equiv V/a} \tag{1.12}$$

### 1.4.7 Standard Atmosphere

The Standard Atmosphere gives representative values for pressure, temperature, density, and kinematic viscosity in air as a function of altitude ($z$) above sea level. This information allows for a common reference when comparing aeronautical data variables of interest without having to predict actual atmospheric properties at any particular Earth location and/or at any particular time of day. The values presented come from measurements combined with certain modelling of the atmosphere. As is customary, Figure 1.5 shows these variables ratioed to their sea-level

**Table 1.2** Flight regimes of significance in aerodynamics.

| | |
|---|---|
| $0 < Ma \leq 0.3$ | Density changes can be neglected (incompressible) |
| $0.3 < Ma < 1.0$ | Subsonic (compressible) flow |
| $0.7 \leq Ma \leq 1.4$ (approximately) | Transonic (compressible) flow |
| $1.0 < Ma < 5.0$ | Supersonic (compressible) flow |
| $Ma > 5.0$ (approximately) | Hypersonic (compressible) flow |

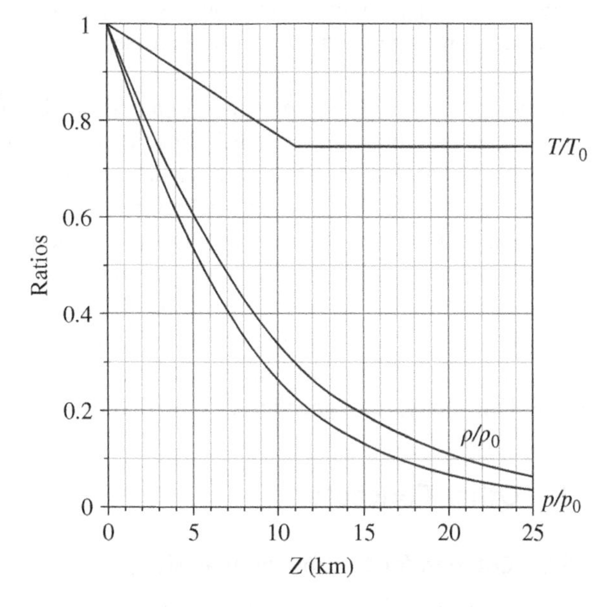


**Figure 1.5** Standard Atmosphere variation for the *ratios* of temperature ($T/T_0$), density ($\rho/\rho_0$), and pressure ($p/p_0$) within the *troposphere* and *stratosphere* as a function of altitude $z$ in km. Here $T_0 = 288.16$ K, $\rho_0 = 1.225$ kg/m$^3$, and $p_0 = 1.013 \times 10^5$ N/m$^2$ at sea level ($z = 0$).

counterparts. Appendix A gives a condensed tabular version of a "1976 US Standard Atmosphere" and several other versions of the same are available on the Internet.

Air molecules are subject to the force of gravity and air pressure is a maximum at the Earth's surface and decreases to zero as we approach space conditions. The differential equation that describes this variation is given from *fluid-statics* as

$$dp/dz = -\rho g \tag{1.13}$$

where $p$ is gas pressure, $\rho$ is gas density, $z$ is elevation, and g is an effective gravity constant. Because atmospheric air temperatures vary with elevation, when substituting $\rho$ in Eq. (1.13) from the prefect gas equation, it is necessary to account for a temperature profile with elevation before integrating it (see Example 1.3 and Problem 1.1). Figure 1.5 depicts standardized atmospheric trends for temperature, density, and pressure ratios as they vary with altitude.

**Example 1.4** In the upper region of the atmosphere called the *stratosphere* (i.e. above $z = 11$ km), the temperature remains relatively constant with height. Integrate the fluid-static relation, Eq. (1.13), for a perfect gas with a constant temperature ($T_s$).

The integration steps follow. $T_s$ is the absolute temperature effective in the stratosphere.

$$\frac{dp}{dz} = -\rho g = -\frac{p}{RT_s}g \tag{E1.4}$$

$$\int_1^2 \frac{dp}{p} = -\int_1^2 \frac{g}{RT_s}dz$$

$$\frac{p_2}{p_1} = \frac{\rho_2}{\rho_1} = \exp\left(-\frac{g(z_2 - z_1)}{RT_s}\right)$$

Since the temperature stays constant in the stratosphere, the density decreases in proportionation to the decrease in pressure with elevation. The pressure and density also decrease with altitude in the *troposphere* which is the region below 10 km but not in such an exponential manner. The decrease in the density with elevation has implications for aircraft during long flights as the aircraft's fuel consumption significantly decreases its weight.

Many aircraft *altitude indicators* or *altimeters* have been barometric, meaning they measure altitude by sensing air pressure. At their flight altitudes, they compare measured static atmospheric pressures with the pressure in an evacuated enclosure connected through a pressure-sensitive unit via an external tap. Such readings need to be then related to altitude through some common or standard model of the atmosphere. Newer types of altimeters utilize electromagnetic wave reflections to more accurately measure height relative to the ground; these include radio altimeters and light detection and ranging (LIDAR). Also noteworthy, since air's viscosity is a function of temperature, the thickness of boundary layers around the wings will decrease at the higher altitudes. Air's *kinematic viscosity* ($\nu$) gently increases with altitude and is often tabulated with the other three in Standard Atmosphere tables, see Appendix A (recall that the units of this property ratio are $m^2/s$ or $ft^2/sec$).

## 1.5 Other Concepts

### 1.5.1 Criterion for Continuum Analysis

The curves depicted in Figure 1.5 (and in the rest of the book) are based on the atmospheric density behaving as *a continuum* which means that air molecules act as vast agglomerations of particles and not as individual entities. Furthermore, these particles are not stationary but move randomly in proportion to their temperature. The scientific description of a continuum requires that the average distance between molecules undergoing collisions be extremely small compared to any characteristic dimension of the immersed object under study. This way we observe only bulk behavior in the gaseous medium. It is safe to assume that all atmospheric flows we encounter below extreme hypersonic regimes will behave under such a continuum so that the gas density exists and has a definable value.

### 1.5.2 The Most Advantageous Viscous Envelope

As shown in Figure 1.6, the Kutta condition along with the presence of thin attached boundary layers are necessary ingredients for an "aerodynamic sweet spot" or a most advantageous viscous envelope – an operating condition wherein subsonic airfoils develop lift while exhibiting low drag. Being managed by a properly rounded leading edge and a sharp trailing edge, viscous effects between the free stream flow and the airfoil suitably configure the developing streamlines over the airfoil as shown on the right side of Figure 1.6. When the resulting boundary layers remain attached and sufficiently thin, aerodynamic conditions outside them may be treated as irrotational

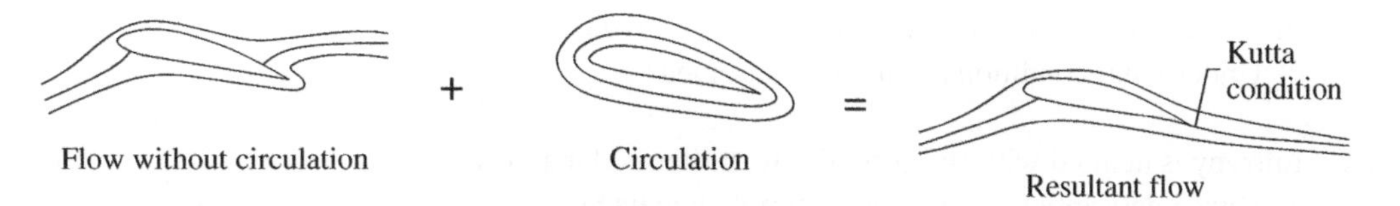


**Figure 1.6** Because viscous effects at the sharp trailing edge produce a stagnation point, the necessary amount of circulation is generated around the airfoil to yield the flow pattern depicted on the right.

flows surrounding an airfoil of equal physical dimensions. We will model *airfoil lift* with a *bound vortex* that introduces circulation into the *uniform free stream flow* as depicted in this figure and show how such fluid *circulation* may be isolated. In finite wings circulation also induces a *drag-due-to-lift* component which will be similarly modelled with a series of *truncated or horseshoe vortices* attached at the airfoil's span.

In supersonic and hypersonic flows, additional phenomena arise and we will focus on the thin flat plate as the most useful airfoil profile. Except for the management of shocks and supersonic expansions, supersonic flows on flat surfaces are relatively simpler to analyze than subsonic flows as will be evident in Chapters 8 and 9.

### 1.5.3 Frame of Reference

In order to avoid unsteady conditions, when we analyze the flow that surrounds a wing, we will "ride" on the flying object as shown in Figure 1.7. This means that in our "control volume analysis," covered beginning in Chapter 2, we will always have steady flows. This type of analysis helps focus on forces and related parameters acting on the aircraft or other object under study and will be part of our default approach. In aerodynamics, there are some topics where unsteady flows are of importance but these belong in more advanced treatments.

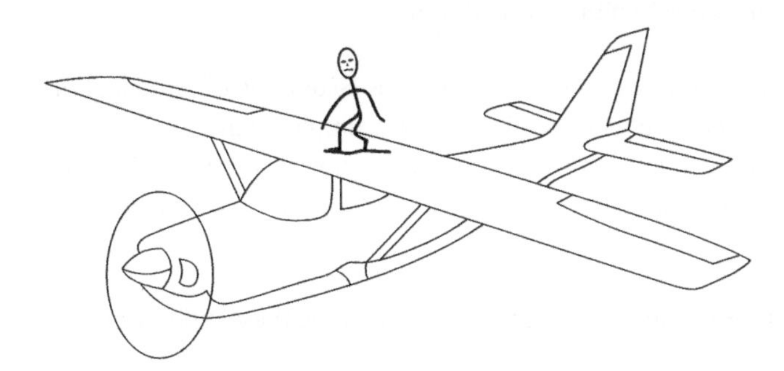

**Figure 1.7** Reference frame taken relative to the flying craft.

## Review Questions

Note: These questions are intended to refresh your background.

**1.1** In fluid mechanics, we introduce the so-called *conservation laws* for mass, momentum, and energy that govern aerodynamic flows.
   a) Under what conditions is mass conserved?

b) Under what conditions is momentum conserved?
c) Under what conditions is energy conserved?

**1.2** Entropy is defined with the second law of thermodynamics.
a) Give a description of entropy in simple language.
b) Is entropy conserved under heat transfer conditions?
c) Is entropy conserved where there are noticeable viscous shearing effects?
d) Is entropy conserved in supersonic flow across a shock?

**1.3** The perfect gas equation of state is used extensively in aerodynamics.
a) Under what temperature and pressure restrictions does it apply?
b) Can it be used with gases other than air?
c) Is it valid only with absolute temperatures and pressures?

**1.4** Gas pressures deliver *normal forces* to exposed surfaces.
a) Describe what is meant by static (or hydrostatic) pressure.
b) Describe what is meant by dynamic pressure.
c) Why is the sum of the above two pressures called the stagnation pressure?

**1.5** Isaac Newton's name is attached to several different phenomena of importance in aerodynamics.
a) Besides his three physical laws, can you name two other phenomena that also carry his name?
b) Besides the momentum concept, what property does Newton's second law introduce?
c) What property is introduced in Newton's law of gravitation?

**1.6** For fluid flows, there are two complementary types of analysis.
a) Explain what is meant by the control mass description.
b) Explain what is meant by the control volume description.

**1.7** The "relation among properties" or $0^2$-*law* of thermodynamics is seldom listed as such, but without it, the entire structure of the field would collapse. For example, in gases where the perfect gas law does not apply, we can still postulate a relation between pressure, temperature, and density (be it analytical or tabular). Can you think of another example of relevance to aerodynamics?

**1.8** What is meant by adiabatic, isothermal, and isobaric processes? How do these differ from the regular ideal reversible processes?

**1.9** What are some flow effects that cause processes to be irreversible?

**1.10** What is the difference between energy and enthalpy and why is the latter more useful in control volume descriptions?

**1.11** For prefect gases, the specific (or per unit mass) internal energy and enthalpy are functions of what variables?

## Problems

**1.1** The lower region of the atmosphere up to about $z = 11$ km is called the troposphere and here the temperature decreases approximately linearly with height. Integrate the fluid-static Eq. (1.13) for such a varying-temperature perfect gas using $T = T_0 - \beta z$, where the "lapse rate" is the constant $\beta = 0.0065$ K/m.

**1.2** Find the density of seawater and calculate the depth at which the pressure doubles from standard sea level. Note in air, the pressure at an elevation of about $z = 5700$ m is half of that at sea level.

**1.3** Confirm the perfect gas the relation $c_p = c_v + R$ for nitrogen. At standard conditions, the tables give the following information: $c_p = 1{,}040$J/kg-K, $c_v = 741$ J/kg-K, and molecular mass = 28.02.

a) How well does this relation represent the data?
b) Calculate the value $\gamma = c_p / c_v$.

**1.4** a) Using $T/\rho^{\gamma-1}$= constant for an *isentropic process*, calculate the ratio of densities when the absolute temperature ratio increases by a factor of two.
b) What will be the ratio of the corresponding absolute pressures for a perfect gas?

**1.5** The equation for fluid statics may be integrated with knowledge of the relation between temperature and elevation. Taking $T_0 = 15$ °C, $z = 11.1$ km, and $p_0 = 1.0132 \times 10^5$ N/m$^2$, compare the value from Figure 1.5 (or Appendix A) with calculated values for pressure resulting from:

a) An isothermal atmosphere as given in Example 1.3 where $T_s = 216.66$ K.
b) An atmosphere where the temperature decreases linearly with height, i.e. $T = T_0 - \beta z$ where $\beta = 0.0065$ K/m and g = 9.81 m/s$^2$. Here $p/p_0 = (1 - \beta z/T_0)^{\frac{g}{\beta R}}$.
c) Interpret your results. $z = 11.1$ km is said to be the location where the *troposphere* meets the *stratosphere*.

## Glossary of Terms and Symbols

|  |  |
|---|---|
| Aerodynamic center, wing | (a.c.) (chord location where changes of lift take place) |
| Area, area vector | ($A$, $\vec{\mathbb{A}}$) |
| Airfoil (wing) | (lift-producing component) |
| Airplane efficiency factor | ($e$) |
| Altitude above sea level, (elevation) | ($z$) |
| Aspect ratio (fineness ratio) | ($\boldsymbol{AR}$) |
| Angle of attack, angle | ($\alpha$, $\theta$) |
| Angle of attack, zero lift | ($\alpha_{\ell 0}$) |
| Boundary layer thickness | ($\delta$) |
| Camber (mean line) | (nonsymmetric, subsonic only) |
| Circulation | ($\Gamma$) |
| Chord, airfoil | ($c$) |
| Critical Mach number | ($Ma_{cr}$) |
| Critical pressure coefficient | ($C_{p,cr}$) |
| Cruise mode | (no accelerations, steady flow) |
| Density (gas) | ($\rho$) |
| Drag | ($d$, $D$) |
| Drag coefficient | ($c_d$, $C_D$) |
| Drag, induced | ($D_i$) |
| Downwash velocity | ($w$) |
| Elliptical planform | ($S_e$, see Chapter 6) |
| Energy, total (specific) | ($e$) |
| Enthalpy (specific) | ($h$) |
| Entropy (specific) | ($s$) |
| Equivalent airspeed | ($EAS$) |
| Flat plate | (symmetric thin airfoil) |
| Flap angle | ($\eta$) |
| Force coefficient | ($C_F$) |
| Free stream velocity | ($V_\infty$) |
| Gas constant | ($R$) |
| Gravitation constant | ($g$) |
| Hypersonic flow | ($Ma > 5.0$) |
| Horseshoe vortex | (see Chapter 6) |
| Ideal flow | (reversible isentropic) |
| Induced drag coefficient | ($C_{Di}$) |
| Internal thermal energy | ($u$) |
| Kinematic viscosity | ($\nu$) |
| Kinetic energy | ($KE$) |
| Kutta condition | (rear-point stagnation condition, subsonic) |
| Laminar flow | (no crossflow mixing) |
| Lift | ($\ell$, $L$) |
| Lift coefficient | ($c_\ell$ , $C_L$) |
| Lift coefficient slope, sectional | ($a_0$ or $c_{\ell\alpha}$) |
| Lift coefficient slope, total | ($a$ or $C_{\ell\alpha}$) |
| Mach number | ($Ma$) |

| | |
|---|---|
| Maximum lift coefficient (sectional) | ($c_{\ell max}$) |
| Maximum thickness-to-chord ratio | ($t_m/c$) |
| Mean aerodynamic chord | (m.a.c.) |
| Molecular mass | (M. M.) |
| Moment coefficient | ($c_m$, $C_M$) |
| Nested chord length | ($c_0$) |
| Oblique shock angle | ($\theta$) |
| Oblique shock deflection angle | ($\delta$) |
| One-dimensional flow | ($x$) |
| Pitching moment | ($m$, $M$) |
| Planform area | ($S$) |
| Polytropic process exponent | ($n$) |
| Potential function | ($\phi$) |
| Potential function (sonic) | ($\varphi$) |
| Potential energy | ($PE$) |
| Power assist angle (lift) | ($\beta$) |
| Prandtl–Meyer angle | ($\nu$) |
| Pressure | ($p$) |
| Pressure coefficient | ($C_p$) |
| Reynolds number | ($Re$) |
| Root chord tapered wing | ($c_r$) |
| Shear stress | ($\tau$) |
| Span, wing | ($b$) |
| Specific property | (per unit mass) |
| Specific heats | ($c_v$,$c_p$) |
| Specific-heat ratio | ($\gamma$) |
| Specific volume | ($v \equiv 1/\rho$) |
| Speed of sound | ($a$) |
| Stall speed | $V_{stall}$ |
| Subsonic flow | ($Ma < 1.0$) |
| Supersonic flow | ($1.0 < Ma \leq 5.0$) |
| Standard atmospheric density at sea level | ($\rho_0$) |
| Standard atmospheric pressure at sea level | ($p_0$) |
| Standard atmospheric temperature at sea level | ($T_0$) |
| Stream function | ($\psi$) |
| Sweepback angle, wing | ($\Lambda$) |
| Symmetric airfoil | (no camber) |
| Taper ratio (tapered wing) | ($\lambda = c_{t/}c_r$) |
| Temperature | ($T$) |
| Thickness ratio, wing (at maximum thickness) | ($t_m/c$) |
| Three-dimensional (3D) ($x$, $y$, $z$) | (upper-case symbols) |
| True airspeed | ($TAS$) |
| Thrust | ($T$) |

| | |
|---|---|
| Tip chord (tapered wing) | ($c_t$) |
| Transonic flow | ($0.7 \leq Ma \leq 1.2$ [approximately]) |
| Turbulent flow | (vigorous crossflow mixing) |
| Twist | (different $\alpha$ along wing span) |
| Two-dimensional (2D) ($x$, $y$) or ($r$, $\theta$) | (lower-case symbols) |
| Unit conversion factors | $g_c = 32.174$ lbm - ft/lbf - sec$^2$ or 9.807kg - m/kgf - sec$^2$ |
| | $J = 778$ ft - lbf/Btu or 4186 joule/kcal |
| Velocity components (Cartesian) | ($u$, $v$, $w$), |
| Velocity components (2D Polar) | ($V_r$, $V_\theta$), |
| Velocity, free stream | ($V_\infty$) |
| Viscosity coefficient | ($\mu$) |
| Volume | ($V$, $\mathbb{V}$) |
| Wash out | (wing twist near tip) |
| Weight | ($W$) |
| $x$-coordinate | length, parallel to $V_\infty$-direction |
| $y$-coordinate (2D) | crossflow |
| $y$-coordinate (3D) | airfoil span, width of wing |
| $z$-coordinate (3D) | axis perpendicular to $V_\infty$–$y$ plane, height |

# 2

# Fluid Dynamic Fundamentals

## 2.1 Introduction

In this chapter, we present the equations that govern fluid flow in regions where viscous effects may be neglected. This enables the use of ideal descriptions but only after such flows have been modified by the Kutta condition as shown in Figure 1.6. Vector calculus notation allows for the compact or "shorthand writing" of these equations, although they will be fully written out as we specialize to one- or two-dimensional flows. A full discussion of *dimensional analysis* is also presented here with association to the dimensionless variables and expressions that capture conventional results without requiring special theorems. A *potential function* ($\phi$) formulation developed for analyzing ideal irrotational flows with a single governing equation using small perturbation theory is given in this chapter. This potential function has been extensively studied and will be needed in Chapters 8–10 where compressible flow effects are discussed. We will begin our study of incompressible aerodynamics in Chapter 5 where elements of *thin airfoil theory* will be introduced compatible with our default approach as outlined in Chapter 1 when the airfoil is properly streamlined.

## 2.2 Objectives

After successfully completing this chapter, you should be able to:

1) Define the potential function ($\phi$), its significance in irrotational/inviscid flow, and how it is related to the stream function ($\psi$).
2) Write the governing equations for continuity, momentum, and energy in their differential form (in vector notation) and identify the physical meaning of each term.
3) Define the lift and drag forces and pitching moments as they relate to an airfoil.
4) Describe how "dimensionless coefficients" differ from the individual variables they contain and why when properly defined, they can be more useful.
5) Enumerate significant phenomena that the following non-dimensional groupings relate to: the Reynolds number, the Mach number, the angle of attack, and the aspect ratio (***AR***).
6) Discuss the dimensionless force relation $C_{\vec{F}} = f(\alpha, Ma_\infty, \boldsymbol{AR}, Re_c)$ in terms of how it represents most flows of interest below the hypersonic regime and how it applies to airfoil analyses.
7) Discuss under what conditions the small perturbation equation becomes either Laplace's equation or the wave equation.

*Elements of Aerodynamics: A Concise Introduction to Physical Concepts*, First Edition. Oscar Biblarz.

Companion website: www.wiley.com/go/elementsofaerodynamics

## 2.3 Control Volume Approach

Unlike introductory physics courses that deal with objects of fixed mass, i.e. the *control mass* approach, in fluid mechanics, it is more convenient to examine a *fixed region in space through which a fluid moves*, the dashed region in Figure 2.1, called the *control volume* approach. This approach is especially useful in aerodynamics because we regularly examine air flows from the point of view of the airplane, i.e. around the flying object (as in Figure 1.7). The transformation from the control mass to the control volume approach is elegantly done through a relation known as the *Reynolds transport theorem* (see Zucker and Biblarz 2020) and is also referred to as a Lagrangian to Eulerian transformation. In *gas dynamics*, the fundamental equations for the control volume are commonly introduced in integral form, but in *aerodynamics*, it is more convenient to use them in differential form. As we show, the *divergence theorem* of vector calculus can be used to transform control surface integrals into control volume integrals so as to blend them together.

### 2.3.1 Notes on Vector Calculus and the Del-Operator $\nabla(\,)$

We use the del-vector $\nabla(\,)$ operator as convenient shorthand for writing our governing equations. The three-dimensional form of this operator is given in Eq. (2.1) in Cartesian form.

$$\nabla(\,) = \vec{i}\frac{\partial(\,)}{\partial x} + \vec{j}\frac{\partial(\,)}{\partial y} + \vec{k}\frac{\partial(\,)}{\partial z} \tag{2.1}$$

Here, $(\vec{i}, \vec{j}, \vec{k})$ are unit vectors that specify three orthogonal space directions. Some relevant mathematical properties using this operator are also summarized below. To represent a velocity

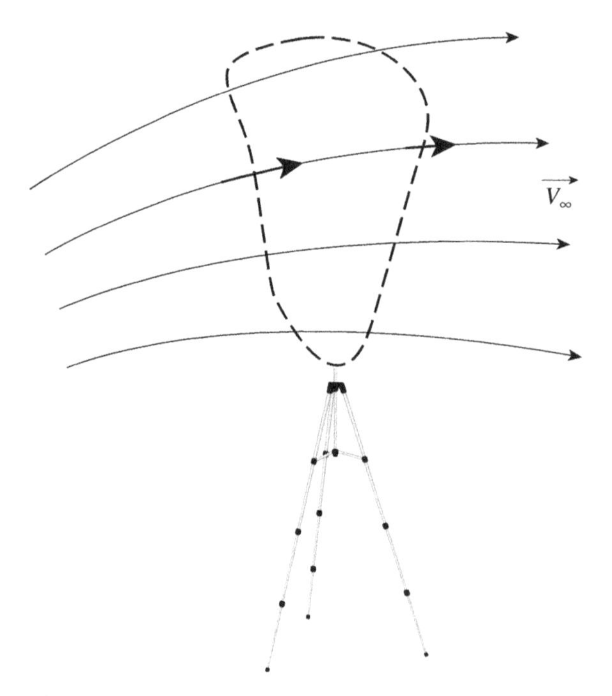


**Figure 2.1** Control volume (the dashed region) fixed in space with flow streaming through it.

vector in vorticity-free flows, we may use the irrotational *potential function* $\phi$ with a gradient (or steepest descend) written as

$$\text{Velocity vector: } \vec{V} = \text{grad}\phi = \nabla\phi \tag{2.2}$$

Two important del-operations with the vector $\vec{V}$ (i.e. the gas velocity) follow:

$$\text{Divergence (fluid source or sink): div}\vec{V} = \nabla \cdot \vec{V} \tag{2.3}$$

$$\text{Curl (fluid rotation): curl}\vec{V} = \nabla \text{x} \vec{V} \tag{2.4}$$

Using Eq. (2.4), the vector relation $\nabla \times \nabla\phi = 0$ verifies that the velocity potential is only defined in irrotational flows. In order to satisfy mass conservation, the flow must be non-divergent or $\nabla \cdot \nabla\phi = \nabla^2\phi = 0$ which will be shown to be the mass continuity equation in incompressible flows. The operator $\nabla^2(\ )$ is known as the Laplacian.

We also use the two-dimensional *stream function* $\psi$ which physically represents lines of fluid flow because streamlines are tangent to the velocity vector (automatically satisfying continuity in incompressible flows) at every point. The stream function is defined in two-dimensional Cartesian flows (but also definable in axisymmetric flows) as

$$u = \frac{\partial\psi}{\partial y} \text{ and } v = -\frac{\partial\psi}{\partial x} \tag{2.5}$$

By definition, along a streamline $d\psi = -vdx + udy = 0$ because $\psi$ cannot change. Also, along streamlines their slope $dy/dx = u/v$. From Eq. (2.4), in irrotational flows, $\nabla \times \vec{V} = 0$ which together with the relations in Eq. (2.5) leads to $\nabla^2\psi = 0$; this means that, akin to the potential function, $\psi$ is also governed by Laplace's equation but with different boundary conditions. Moreover, it will be shown that $\psi$-lines and $\phi$-lines form *perpendicular grids*, meaning that streamlines are perpendicular to equipotential lines. Flows described with either of these functions may be added to each other because Laplace's equation is a linear differential equation – this is the so-called principle of *superposition of elementary flows* which is used to represent complex flows from simpler ones. As will be elaborated in Chapter 3, typical two-dimensional elementary flows used in aerodynamics consist of:

i) uniform flow
ii) vortex flow
iii) sources and sinks

We will make extensive use of the first two in our *thin airfoil theory* formulations of Chapters 5 and 6.

### 2.3.2 Conservation of Mass (Continuity Equation)

Mass continuity is one of the foundational equations in fluid flow. We introduce here the integral form of this equation for the control volume (equation (2.25) in Zucker and Biblarz 2020) which may be written as follows

$$\frac{\partial}{\partial t}\int_{\text{cv}} \rho d\mathbb{V} + \int_{\text{cs}} \rho\left(\vec{V} \cdot d\vec{\mathbb{A}}\right) = 0$$

The first term is a volume integral and the second integral is an area integral in which the area vector is in normal direction to the surface and one that, after being transformed into a volume integral through the use of the *divergence theorem*, is merged with the first integral. This results in

$$\int_{cv} \left[\frac{\partial \rho}{\partial t} + \nabla \cdot \left(\rho \vec{V}\right)\right] d\mathbb{V} = 0 \tag{2.6}$$

Since the argument of the volume integral must itself be zero,

$$\boxed{\frac{\partial \rho}{\partial t} + \nabla \cdot \left(\rho \vec{V}\right) = 0} \tag{2.7}$$

Steady and incompressible flow is an important application in aerodynamics, and here the equation simplifies to

$$\nabla \cdot \vec{V} = 0 \; [\rho = \text{constant}] \tag{2.8}$$

Equation (2.8) states that the divergence of the velocity vector must vanish in steady incompressible flows. Physically, this means that there can be no fluid sources or sinks inside our chosen control volume or that there can be no difference between net outflows and net inflows through a control volume because mass must be conserved.

**Example 2.1** Show that the elementary flow called *uniform flow* satisfies the continuity equation for incompressible flow but that another elementary flow, aptly called *source flow*, does not. [Note: related flow diagrams are shown in Chapter 3, here we only want to present some mathematical details.]

We are concerned with flows that satisfy Eq. (2.8). In uniform flow, the velocity is constant, and since the del-operator takes derivatives of the velocity, vector we could simply reason that any derivative of a constant is zero, but we here shall do it more formally. The relevant potential functions are typically written as

$$\text{Uniform flow: } \phi = V_\infty x$$

where $x$ is the Cartesian direction of the velocity vector.

$$\text{Source flow: } \phi = (\lambda/2\pi) \ln r$$

where $\lambda$ is the source strength and $r$ a radial direction from the source.

For *uniform flow,* $u = \phi_x = V_\infty$ so that $\nabla \cdot \vec{V} = \dfrac{d(V_\infty)}{dx} = 0$.

For *source flow,* the streamlines issue radially from a central location inside the control volume. Here it is more convenient to work with cylindrical coordinates for which the del-operator is written as

$$\nabla() = \vec{e}_r \frac{\partial()}{\partial r} + \frac{\vec{e}_\theta}{r}\frac{\partial()}{\partial \theta} + \vec{e}_z \frac{\partial()}{\partial z}$$

A two-dimensional velocity component for source flow is only radial in cylindrical coordinates, $v_r = \lambda/2\pi r$. Finally, we find $\nabla \cdot \vec{V} = -\dfrac{\lambda}{2\pi r^2} \neq 0$ for ordinary values of the radius. In order to satisfy continuity, we could add to this source a nearby sink of equal strength and define a control surface that encloses both.

### 2.3.3 Momentum Equation (Without Shearing or Gravity Forces)

We continue now with components in integral form for the momentum equation for the control volume, (equation (3.41) in Zucker and Biblarz 2020), retaining only pressure forces. We then transform the surface integral on the left side of the equal sign into a volume integral as before and merge it with the locally unsteady term to obtain equation (2.10)

$$\frac{\partial}{\partial t}\int_{cv}\vec{V}\rho d\mathbb{V} + \int_{cs}\vec{V}\rho\left(\vec{V}\cdot d\vec{\mathbb{A}}\right) = \int_{cs} p d\vec{\mathbb{A}} \tag{2.9}$$

$$\boxed{\rho\frac{\partial\vec{V}}{\partial t} + \rho\left(\vec{V}\cdot\nabla\right)\vec{V} = -\nabla p} \tag{2.10}$$

Equation (2.10) has been restricted to cases where body forces and viscous forces may be neglected in order to correspond to our stated ideal flow conditions. The force of gravity is a body that affects individual gas molecules and is reflected in the local atmospheric pressure, and viscous forces are unimportant outside the boundary layers. The last term in Eq. (2.10) represents the *pressure force gradient* effective around immersed objects; the gradient operation stems from the divergence theorem when applied to the area integral of the pressure which is a scalar term.

### 2.3.4 Energy

We can similarly manipulate the energy equation [equation (2.35) in Zucker and Biblarz 2020] with kinetic and internal energies included but without potential energies. The transformation to differential control volume form of the energy equation in *adiabatic flows with no shaft work and with no viscous work* but including *flow work* as per Eq. (1.10), after using the divergence theorem, becomes

$$\boxed{\frac{\partial}{\partial t}\left[\rho\left(\frac{V^2}{2} + u\right)\right] + \nabla\cdot\left[\rho\vec{V}\left(\frac{V^2}{2} + h\right)\right] = 0} \tag{2.11}$$

where $u$ and $h$ represent the fluid's specific *internal thermal energy* and the *enthalpy*, respectively, and $V^2/2$ is kinetic energy per unit mass. This equation is not used with incompressible flows but necessary with compressible flows. At very high Mach numbers, in hypersonic situations, thermal conduction and viscous energy dissipation terms need to be included because these flows are highly non-adiabatic.

The foregoing differential equations form the necessary background that relates to ideal-flow fundamentals. Equation (2.10) is called Euler's equation, and we will use its time-independent form in Chapter 4. In Chapter 7, we add to Euler's equation a viscous-force term in order to examine conditions inside the boundary layers – together with Eq. (2.8), this set is known as the Navier–Stokes equations. As written, Eqs (2.7) and (2.10) apply in all Mach number ranges of interest. In Chapter 8, we use the steady state form of Eq. (2.11). Lastly, when the velocity is expressed in terms of the *potential function* from Eq. (2.2), Euler's equation together with the continuity equation and the isentropic pressure-density relation, namely Eq. (1.7),

$$p\rho^{-n} = \text{constant} \quad \text{with } n = \gamma$$

form the basis for *small perturbation theory*. This theory is discussed in Section 2.6 and used in the compressible flow descriptions developed in Chapters 9 and 10.

## 2.4 Lift, Drag, and Pitching Moment

The resultant aerodynamic force acting on an airfoil section is divided into two components relative to the direction of the free stream velocity, one parallel to it called the *drag* and another perpendicular called the *lift*. These are depicted in Figure 2.2. The distribution of pressure forces that generate this total force also produces a nose-down (counter-clockwise) moment or *pitching moment* about a lifting airfoil's leading edge as will be seen in Figure 2.4

In the notation and formulations listed in the following and throughout this book, we use upper case symbols for total forces and lower case symbols for sectional or per unit span forces.

a) Lift ($L$, $\ell$). Fluid dynamic lift produced entirely by pressure distributions around the airfoil when *circulation* is present. Under optimal lift, the Kutta condition is satisfied and the flow separates only beyond the airfoil's sharp trailing edge. On an aircraft cruising in level flight, the total lift force must equal its weight (as depicted in Figure 1.1).
b) Drag ($D$, $d$). Drag is the net aerodynamic force acting against the motion of the aircraft. The propulsion system's main role during cruise conditions is to balance the drag force but additional propulsion power needs arise during takeoff, landing, and maneuvering. There are several contributions to the drag that include pressure and shearing forces and these together with the "drag due to lift" will be detailed in Chapters 6, 7, and 8.
c) Pitching Moment ($M$, $m$). A pitching moment results from non-symmetrical distributions of the pressure forces acting on a lifting wing. It is a moment force about the lateral or $y$-axis of the aircraft. A positive moment is defined to cause rotation in the nose-up direction. There are other moment forces that act on an aircraft but these do not originate at the lift-drag force plane shown in Figure 2.2.

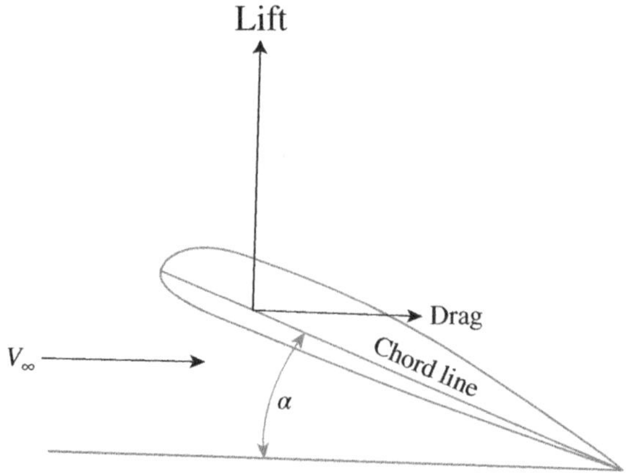

**Figure 2.2** Resultant pressure forces on an airfoil profile section at angle of attack ($\alpha$).

## 2.5 Dimensional Analysis

In order to examine forces and moments acting on an airfoil in a somewhat more general manner, a set of non-dimensional ratios have been developed with a technique called "dimensional analysis" which is in part based on the concept that formulations of natural phenomena should be independent of physical units. Dimensional analysis has been much influenced by experimental observations, scientific tendencies, and the strategic merging and generalizing of these.

In order to examine the *surface forces* that develop from the relative motion between a wing and the medium in which it is immersed, we begin our discussion with a description of a simple wing configuration – one that is thin, flat, and rectangular. From numerous studies, the parameters most relevant under these conditions may be enumerated as follows:

$$\vec{F} = \mathrm{f}(\rho_\infty, V_\infty, \mu_\infty, a_\infty, \alpha, c, b) \tag{2.12}$$

In this relation, the net aerodynamic force vector ($\vec{F}$) has an as yet undefined dependence, or "f(...)," on the density ($\rho_\infty$), the relative velocity ($V_\infty$), the viscosity ($\mu_\infty$), and the speed of sound ($a_\infty$) as well as on the wing's angle of attack ($\alpha$), chord length ($c$), and span ($b$). That is, experience teaches us that the resultant surface force on the airfoil depends on certain air flow properties ($\rho_\infty$, $\mu_\infty$, $a_\infty$), on the relative velocity between the airfoil and the medium ($V_\infty$) together with the geometrical orientation of this airfoil with respect to the free stream velocity vector ($\alpha$), and on the wing's planform area configuration ($c$, $b$). The "angle of attack" has been included here as a "geometrical input" to represent the wing's planform elevation above the plane of the velocity vector, thereby introducing the cross-flow variable $z$ (i.e. $\alpha = \sin^{-1}(z/c)$). This angle, however, is already dimensionless and will remain unmodified in our final results, something that also applies to $\gamma$ another dimensionless parameter.

The procedure in dimensional analysis starts by identifying parameters in Eq. (2.12) that are relatively constant in the problem and that as a group contain the necessary set of dimensions – for wings in subsonic flows (and *mechanics* in general), we need to select only three independent *reference parameters* and these are typically $\rho_\infty$, $V_\infty$, and $c$ because this set contains the three *fundamental dimensions* of *mass*, *length*, and *time*. The equation of state adds another, namely, the *temperature* which must be treated as a fundamental dimension as well.

The remainder set of dimensional parameters in Eq. (2.12) $\vec{F}$, $a_\infty$, $b$, and $\mu_\infty$, is made dimensionless by manipulating each one with the chosen set of *reference parameters* so as to appropriately cancel units. The resulting dimensionless force coefficient turns out to be

$$\left(\frac{\vec{F}}{\rho_\infty V_\infty^2 c^2}\right)$$

This form is made more useful by replacing $c^2$with the wing's *planform area* $S = bc$ which has the same units but can represent configurations more general than rectangular planforms. We would also like to have the *dynamic pressure* instead of $\rho_\infty V_\infty^2$ because it appears in Bernoulli's equation as will be shown in Chapter 4 (Eq. (4.5)). The dynamic pressure has the same units as the pressure and reads as $q_\infty = ½\rho_\infty V_\infty^2$, and this requires inserting the factor ½ which being a pure number does not affect the non-dimensionality of the equation. As already mentioned, the net force being a vector has components for both lift and drag which depend on the angle of attack ($\alpha$) of the airfoil and other aspects of the air flow. A resulting *force coefficient* may now be written as

$$C_{\vec{F}} = \frac{\vec{F}}{\frac{1}{2}\rho_\infty V_\infty^2 S} = \frac{\vec{F}}{q_\infty S} \tag{2.13}$$

Next we non-dimensionalize the speed of sound $a_\infty$ which has dimensions that directly match $V_\infty$ so that their ratio is sufficient to eliminate all units. There already exists a well-known form of such a ratio, namely, the *Mach number* given by

$$Ma_\infty = \frac{V_\infty}{a_\infty} \tag{2.14}$$

As stated earlier, we wish to work with the planform area $S$ so that for a rectangular wing, the wing span ($b$) which is non-dimensionalized by the chord ($c$) may also be written as

$$\boldsymbol{AR} = \frac{b}{c} = \frac{b^2}{S} \tag{2.15}$$

This form is called the *aspect ratio*, and similar to the Mach number, this is the conventional form used for these parameters. In non-rectangular wings, the ratio given after the second equal sign in Eq. (2.15) generalizes the definition of this coefficient because it may now accommodate a new dimensionless number called the *taper ratio* ($\lambda$). A rectangular wing has a taper ratio of 1.0 and a pointed tip wing has no taper ratio, with trapezoidal planforms having values in between.

Finally, we need to non-dimensionalize the viscosity and not surprisingly, its inverse ratio already exists, known as the Reynolds number based on chord length $c$,

$$Re_c = \frac{\rho_\infty V_\infty c}{\mu_\infty} \tag{2.16}$$

These last three ratios typically remain constant under many flow situations of interest. We can now write a dimensionless form of Eq. (2.12) as

$$\boxed{C_{\vec{F}} = \mathrm{f}(\alpha, Ma_\infty, \boldsymbol{AR}, Re_c)} \tag{2.17}$$

None of the parameters in Eq. (2.17) carries units and the force coefficient is shown with the most conventional dimensionless form of the other coefficients.

**Example 2.2** Verify that the given ratios for the Reynolds number and the force coefficient $C_{\vec{F}}$ are dimensionless.

This can be done in either the SI or EE system of units. We will do it in the SI system for the inverse $Re_c$ and leave the verification of $C_{\vec{F}}$ up to the reader. The units of viscosity are [N·s/m$^2$ or kg/m·s] from Section 1.4 and those for our three reference parameters ($\rho_\infty$, $V_\infty$, and $c$) are, respectively, [kg/m$^3$, m/s, and m] from Section 1.3. We will work with the inverse form of Eq. (2.16) so the unit ratios become

$$\left[\frac{\mathrm{kg/m\cdot s}}{\mathrm{kg/m^3\cdot m/s\cdot m}}\right] = \left[\frac{\mathrm{kg\cdot m^3\cdot s}}{\mathrm{kg\cdot m^3\cdot s}}\right] \equiv \text{dimensionless}$$

In many aeronautical situations, it becomes necessary to introduce additional parameters, for example: in Chapters 5, we discuss *airfoil camber* that adds a negative dimensionless constant which is the angle of zero lift ($\alpha_{\ell 0}$), and in Chapter 9 we introduce another geometry parameter, namely, the *airfoil sweepback angle* ($\Lambda$) which is of importance only in compressible flows. A dimensionless factor labelled $t_m/c$ representing the maximum airfoil thickness-to-chord ratio is also included. Having already mentioned the taper ratio, we now have four new parameters to be added to our master equation which may be written as

$$C_{\vec{F}} = \mathrm{f}(\alpha, Ma_\infty, \boldsymbol{AR}, Re_c, \alpha_{\ell 0}, \Lambda, t_m/c, \lambda) \tag{2.18}$$

One thing that is implicit in the equations above but which bears mention is that the functional dependance has been left unresolved in transforming Eq. (2.12) to Eqs. (2.17) and (2.18). What has changed is the number of factors in the problem – instead of postulating relations between eight parameters in the original equation we are now concerned with only five dimensionless groups in Eq. (2.17) [nine in its expanded version Eq. (2.18)]. Typically, these are more than sufficient and by using conventional dimensionless grouping explicitly the most common. It should also be pointed out that under many conditions $C_{\vec{F}}$ need not depend on the entire group simultaneously (i.e. incompressible flows do not depend on Mach number and airfoils do not need sweepback). Moreover, we do not anticipate $C_{\vec{F}}$ to depend on many others. There are additional advantages in using dimensionless ratios because they are not only independent of any system of units but may also carry a broader meaning – for example, the various aerodynamics regimes described in Chapter 1 are solely identified by their Mach number, and a boundary layer's transition to turbulence can be shown to be primarily a function of the Reynolds number. Even though airfoil thickness and thickness distribution may also affect $C_{\vec{F}}$, under our *default assumption* of *thin airfoils*, these dependencies are treated as negligible.

A follow-on task involves establishing the functional relation relevant to the above formulation. Science has two synergistic approaches, one experimental and the other analytical, and fluid mechanics uses both extensively. Properly instrumented wind tunnels are needed for testing full-size airfoils or more often their smaller prototypes, and for the latter, dimensional analysis is the key to *modelling* and *similitude* (see Streeter and Wylie 1985). These two disciplines are utilized to arrive at $Re_c$ and $Ma_\infty$ values since they govern a prototype's size and are also needed to establish testing parameters. But well instrumented, large wind tunnel facilities are rare and expensive to run. For supersonic and hypersonic flows, most wind tunnels presently lag modern analytical tools that work with today's powerful computers, often having a cost advantage over experiments.

### 2.5.1 Incompressible Flow Results

For a common and important range of Mach numbers, $Ma_\infty < 0.3$ in air, and at angles of attack where the boundary layers remain attached on the airfoil ($-15° \leq \alpha \leq 15°$), symmetric and unswept wings are well described by a subset of the groupings in Eq. (2.17). Let $C_L$ and $C_D$ be a *finite wing's* total lift and drag coefficients, respectively, and $c_\ell$ and $c_d$ their corresponding sectional or two-dimensional counterparts. The lower-case symbols are per-unit span quantities that may vary along the span direction of the wing. Under these conditions, we may write

$$\text{3D } C_L = \frac{L}{q_\infty S} = \mathrm{f}_L(\alpha, \boldsymbol{AR}) \tag{2.19}$$

$$\text{2D } c_\ell = \frac{\ell}{q_\infty c} = \mathrm{f}_\ell(\alpha) \tag{2.20}$$

$$\text{3D } C_D = \frac{D}{q_\infty S} = \mathrm{f}_D(Re_c, \alpha, \boldsymbol{AR}) \tag{2.21}$$

$$\text{2D } c_d = \frac{d}{q_\infty c} = \mathrm{f}_d(Re_c, \alpha) \tag{2.22}$$

$[Ma < 0.3 \text{ in air} \quad (-15° \leq \alpha \leq 15°)]$

We should note here that with modern airfoils the unstalled range can exceed 15° because airfoils increase this capability with their extended flaps and slats as discussed in Chapter 11. For this low subsonic range, effects of Mach and Reynolds numbers on lift can be small. Drag, however, will always depend on Reynolds number because of a significant contribution from skin friction.

For the case where lift depends on only one of the original sets of dimensionless groups, in Chapter 5 we develop the functional relation for Eq. (2.20) for symmetric and cambered airfoils and show that it is actually linear. Here, $c_{\ell\alpha}$ represents the lift-curve slope.

$$\text{Symmetric airfoil: } c_\ell = c_{\ell\alpha}(\alpha) \tag{2.23}$$

$$\text{Cambered airfoil: } c_\ell = c_{\ell\alpha}(\alpha - \alpha_{\ell 0}) \tag{2.24}$$

### 2.5.2 Compressibity and Other Effects

The incompressible slope in Eqs (2.23) and (2.24), $c_{\ell\alpha}$, is constant and is known to have a theoretical value of $2\pi$-radians for thin airfoils. For compressible flows, it will be shown in Chapter 9 that rising Mach numbers in both subsonic and supersonic flows increases $c_{\ell\alpha}$. Higher Reynolds numbers actually slightly decrease this slope but the Reynolds number has a more profound effect on drag and on boundary layer separation as discussed in Chapter 7. A small effect arising from airfoil finite thickness (the ratio $t_m/c$) may also change the theoretical lift-curve slope slightly above $2\pi$. For finite wings, increasing the ***AR*** has a large effect on the lift curve because as we will learn in Chapter 6 it lessens the wing's *end effects*, making airfoils more two-dimensional and asymptotically nudging the slope closer to $2\pi$.

Mach number effects are discussed in Chapters 8–10 with subsonic formulations based on extensions of incompressible flow descriptions. With non-rectangular wings, several additional geometrical parameters have to be considered. As shown in Figure 2.3, for straight but tapered wings, we need the taper ratio ($\lambda = c_t/c_r$); in swept wings, we also need the forward and rear sweepback angle ($\Lambda$), both already mentioned; and for a delta wing, we may also need taper and sweep ratios. The planform area $S$ must represent the physical area of each of these geometries. Last but not least, viscous effects affect only the thin boundary layers which as stated are part of our default assumptions, and other aspects of viscous flow can be found in Chapter 7.

As given in this section, the makeup of $C_L$ along with its two-dimensional counterpart $c_\ell$ becomes defective under both compressible flow conditions and also for incompressible flows at high angles

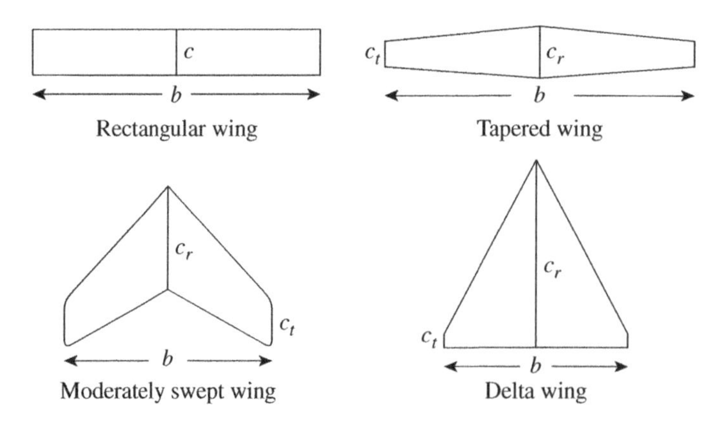


**Figure 2.3** Common wing planform configurations.

of attack. The former arises because dynamic pressure ($q_\infty$) in Eqs. (2.19) and (2.20) contains a $Ma_\infty$-dependance which needs to be factored out as shown in Chapter 9, where we examine Mach number effects on these coefficients. The latter arises because we need to modify the effective chord to include a high $\alpha$-dependence of the lifting planform area and also when there are extended slat/slot units that enhance the chord ($c$), both detailed in Chapter 11.

### 2.5.3 Moment Coefficient

Next we look at pitching moments generated by pressure forces on an airfoil. These moments are important because they influence airplane stability. From mechanics, we know that this moment ($M$) depends on the cross-product of the *force times its moment arm* and that moments contain the same three fundamental units as the force (mass, length, and time). We consider only *pitching moments* here, ones arise from rotation airfoil sections about its span axis, i.e. on the lift free-stream-velocity plane as shown in Figure 2.4. In direct similarity to Eq. (2.12) we write

$$M = \mathrm{f}(\rho_\infty, V_\infty, \mu_\infty, a_\infty, \alpha, c, b) \tag{2.25}$$

Since the moment arm adds an additional length, it may be shown that for incompressible flows at low angles of attack

$$\text{3D } C_M = \frac{M}{q_\infty Sc} = \mathrm{f}_M(\alpha, \boldsymbol{AR}) \tag{2.26}$$

$$\text{2D } c_m = \frac{m}{q_\infty c^2} = \mathrm{f}_m(\alpha) \tag{2.27}$$

In the equations above, we have also assumed a thin, rectangular unswept wing and neglected minor effects arising from the Reynolds number on the moment force. There are two commonly used locations on the airfoil about which we calculate the pitching moment, one is the leading edge and the other is the quarter-point which is also the location of the *aerodynamic center*, or chord location for symmetric airfoils where the moment is independent of angle of attack. By convention, the pitching moments depicted in Figure 2.4 are negative because they result in a "nose down" motion during lift. Discussing Figure 2.4 further, symmetric airfoils need $\alpha > 0°$ to develop lift, whereas cambered airfoils can develop lift at $\alpha = 0°$. This is evident in Figure 2.4 from the length and position of the arrows. For symmetric airfoils, the pitching moment about the aerodynamic

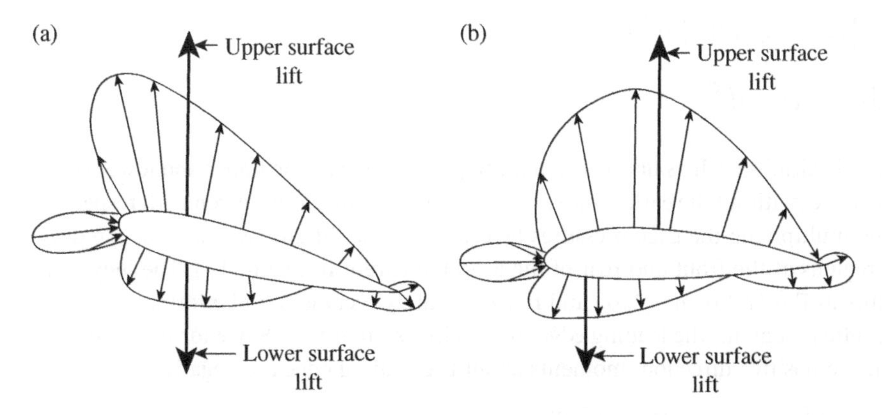


**Figure 2.4** Pressure distributions on our two types of airfoil sections. (a) Symmetric airfoil at $\alpha > 0°$. (b) Cambered airfoil at $\alpha = 0°$.

center (a.c.) is always zero, whereas for the cambered airfoil it is always negative. This figure clearly depicts how pressure forces are solely responsible for the net lift force and shows the resulting effects of the imposed Kutta condition.

The definition of $C_M$ together with its two-dimensional counterpart $c_m$ as presented here becomes defective under conditions other than incompressible flows and low angles of attack because, in addition to compressibility and other effects on the lift force coefficient mentioned earlier, the moment arm perpendicular to the lift turns into a function of angle of attack as discussed in Chapter 11.

### 2.5.4 Pressure Coefficient

Next we develop how lift forces on an airfoil may be calculated from surrounding pressure distributions. In subsonic flows, such distributions are highly non-uniform around the airfoil as shown in Figure 2.4. There, above-atmospheric or positive pressures are shown as arrows *into* the airfoil profile element and below-atmospheric or negative pressures are shown away from the profile. For sufficiently thin airfoils at low angles of attack, we normally neglect the positive pressures at both ends and the angular deviations from the normal of the arrows away from the airfoil. The net pressure force is found by adding (or integrating) the pressures that surround the entire airfoil surface. The subscripts used with the pressure terms below signify $l$ = lower surface, $u$ = upper surface, and$\infty$ = free stream (constant).

$$L = F_z = \oint\!\!\oint (p_l - p_\infty)dxdy - \oint\!\!\oint (p_u - p_\infty)dxdy \tag{2.28}$$

The pressure terms in Eq. (2.28) can be either gauge or absolute and become non-dimensional when divided by the dynamic pressure. In coefficient form, as in Eq. (2.17), the argument inside each integral in Eq. (2.28) becomes the pressure coefficient with the following functional form

$$C_p \equiv \frac{p - p_\infty}{q_\infty} = \mathrm{f}(Re_c, Ma_\infty, \alpha, \boldsymbol{AR}) \tag{2.29}$$

It can be shown that $C_p$ depends on the same dimensionless coefficients as the force. Now, properly non-dimensionalized, the lift coefficients become

$$3\mathrm{D}\ C_L = \oint\!\!\oint (C_{pl} - C_{pu})d\left(\frac{x}{c}\right)d\left(\frac{y}{b}\right) \tag{2.30}$$

$$2\mathrm{D}\ \ c_\ell = \oint (C_{pl} - C_{pu})d\left(\frac{x}{c}\right) \tag{2.31}$$

For thin symmetric airfoils such as flat plates in *incompressible* flows, the upper and lower components of the pressure (without the earlier stated approximations) are closely given with respect to the free stream by multiplying the gage pressures by $\cos(\alpha)$ for the lift and by $\sin(\alpha)$ for the drag (except at small regions at the front and rear of the airfoil apparent in Figure 2.4). The lengths $b$ and $c$ are constants in Eqs (2.30) and (2.31) and reflect a purely rectangular planform.

The moment coefficient about the leading edge follows in a similar way. Since for moment vectors nose up is in the positive direction, moments about the leading edge are negative.

$$3\mathrm{D}\ \ C_{M0} = -\oint\!\!\oint (C_{pl} - C_{pu})\left(\frac{x}{c}\right)d\left(\frac{x}{c}\right)d\left(\frac{y}{b}\right) \tag{2.32}$$

$$\text{2D} \quad c_{m0} = -\oint (C_{pl} - C_{pu}) \left(\frac{x}{c}\right) d\left(\frac{x}{c}\right) \tag{2.33}$$

**Example 2.3** In order to simplify the integrations, we will take a special case where the pressure distributions are constant across the top and bottom surfaces of a symmetric airfoil. Given an upper pressure coefficient value of $C_{pu} = -2.0$ (suction) and a lower coefficient of $C_{pl} = 0.5$ when $\alpha = 5°$, find the resulting force and moment coefficients about the leading edge.

$$c_\ell = \int_0^1 (C_{pl} - C_{pu}) d\left(\frac{x}{c}\right) = 2.5 \cos\left(5^\circ\right) = 2.490$$

$$c_{m0} = -\int_0^1 (C_{pl} - C_{pu}) \left(\frac{x}{c}\right) d\left(\frac{x}{c}\right) = -1.25 \cos\left(5^\circ\right) = -1.245$$

As clearly evident in Figure 2.4, pressure distributions around airfoils in *subsonic flows* are considerably more complicated. In *supersonic flows* over a flat plate airfoil, they are particularly simple being constant above and below the plate and this will be shown in Chapter 9.

## 2.6 Small Perturbation Theory in Steady Compressible Flows

We discuss here the *small perturbation approach* to illustrate how the fundamental equations of Section 2.3 are manipulated in ideal flows to yield a single governing equation. Because the resulting equation is highly nonlinear, in order to arrive at the working formulations presented below and in Chapters 9 and 10, restrictions are introduced to only represent cases where the approaching uniform air velocity is "slightly disturbed" by a slender object like a thin airfoil (hence the label small perturbation). The flow Mach number defined under dimensional analysis ($Ma_\infty \equiv V_\infty / a_\infty$) turns out to be an important parameter for the study of relevant aerodynamic regimes. Necessary assumptions for this theory are summarized next, and Figure 2.5 highlights the regions where irrotational flow may be present around an airfoil.

Assumptions:

i) steady, irrotational, and compressible flow
ii) no body forces, work, or heat transfer
iii) two-dimensional uniform-flow conditions

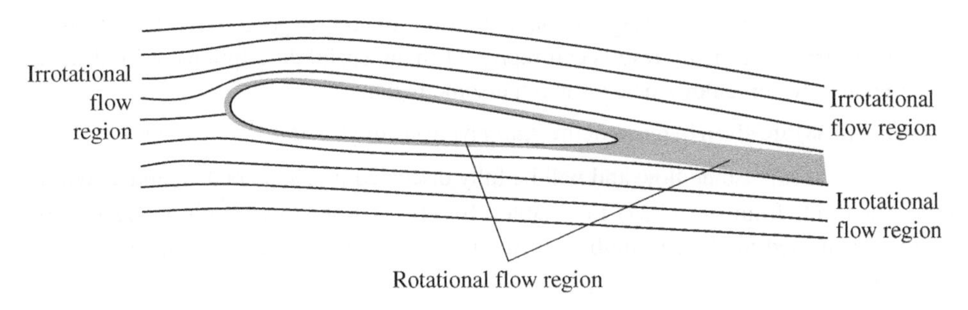


**Figure 2.5** Regions of rotational (shaded) and irrotational flow around an airfoil.

Equations:

$$\nabla \cdot \left(\rho \vec{V}\right) = 0 \tag{2.34}$$

$$\rho\left(\vec{V} \cdot \nabla\right)\vec{V} = -\nabla p \tag{2.35}$$

$$\nabla \times \vec{V} = 0 \quad \text{so that} \quad \vec{V} \equiv \vec{V}_{\infty} + \nabla\phi \tag{2.36}$$

Equation (2.34) is the steady portion of Eqs. (2.7) and (2.35) the steady portion of Eq. (2.10), both written in their fully compressible-flow form. As already stated, irrotationality is synonymous with inviscid isentropic flow and it allows the introduction of the small velocity-perturbation potential function ($\phi$) as a function of $x$ and $y$. After applying the needed isentropic flow relations and much manipulation, the following second-order partial differential equation is obtained (see Anderson 1978, 2017; Bertin and Cummings 2013; Liepmann and Roshko 1957; Saad 1985; Shapiro 1953):

$$\left(1 - Ma_{\infty}^2\right)\phi_{xx} + \phi_{yy} = \frac{Ma_{\infty}^2(\gamma + 1)}{V_{\infty}}\phi_x\phi_{xx} + \cdots \tag{2.37}$$

The potential function $\phi$ as defined has units of $m^2/s$ or $ft^2/sec$ and components of the velocity are written for two-dimensional flows as

$$u = \frac{\partial\phi}{\partial x} \equiv \phi_x \quad \text{and} \quad v = \frac{\partial\phi}{\partial y} \equiv \phi_y.$$

Equation (2.37) is known as the *transonic small perturbation equation* written in terms of the potential function $\phi(x, y)$. Higher-order terms are not shown because we will be able to neglect them. The right-hand-side of this equation is nonlinear but it is only necessary to include it inside the transonic region, and this is discussed more fully in Chapter 10. For purely subsonic or supersonic flows, Eq. (2.37) simplifies to

$$\boxed{\left(1 - Ma_{\infty}^2\right)\phi_{xx} + \phi_{yy} = 0} \tag{2.38}$$

which is a linear partial differential equation. For Mach numbers below one, it contains the well-known Laplace's equation. We have already seen the incompressible form of Laplace's equation, but Eq. (2.38) applies to both incompressible and subsonic-compressible flows, up to Mach numbers of about 0.8 – the factor $\left(1 - Ma_{\infty}^2\right)$ remains as a fixed constant with each approaching Mach number. At supersonic conditions, the sign of the first term flips and we get the well-known *wave equation*. Small perturbation theory results can represent many useful descriptions of isentropic, two-dimensional aerodynamic flows below the hypersonic region.

Solutions to Eq. (2.38) are given in terms of the *multiplying factor* $\left[1 - Ma_{\infty}^2\right]^{-1/2}$ in subsonic flow and $\left[Ma_{\infty}^2 - 1\right]^{-1/2}$ in supersonic flow and will be fully discussed in Chapters 3, 8, and 9. While we do present a result from Eq. (2.37) in Chapter 10, effects from these multiplying factors become more complicated for Mach numbers in the transonic range which can range between $0.7 \leq Ma_{\infty} \leq 1.4$.

The resulting pressure coefficient for two-dimensional *compressible* flows in terms of the potential function given above as Eq. (2.38) follows and may be written as

$$C_p = -\frac{2u}{V_\infty} = -\frac{2}{V_\infty}\frac{\partial \phi}{\partial x} \tag{2.39}$$

## 2.7 Summary

The fundamental equations for ideal fluid flows are transformed from their integral control volume form into differential form. Some benefits from using the potential and stream functions have also been indicated. Dimensionless representations of the most significant aerodynamic parameters are presented in this chapter in their conventional forms but no attempt is made to derive them. Small perturbation theory formulations in terms of the potential function are introduced in this chapter for use in the chapters of compressible flow.

Equations (2.7) and (2.10) written in differential form are significant because they represent conservation of mass and the momentum equation in inviscid flow regions.

$$\boxed{\frac{\partial \rho}{\partial t} + \nabla \cdot \left(\rho \vec{V}\right) = 0} \tag{2.7}$$

$$\boxed{\rho \frac{\partial \vec{V}}{\partial t} + \rho\left(\vec{V} \cdot \nabla\right)\vec{V} = -\nabla p} \tag{2.10}$$

Tables 2.1–2.3 summarize some dimensional analysis results:

**Table 2.1** Summary of 3D dependencies for $\alpha < \pm\, 15°$.

| | ***Ma* < 0.3** | ***Ma* ≥ 0.3** |
|---|---|---|
| $C_L$ | f($\alpha$, $\alpha_{\ell 0}$, ***AR***, $\lambda$, $\Lambda$) | f($Ma_\infty$, $\alpha$, $\alpha_{\ell 0}$, ***AR***, $\lambda$, $\Lambda$) |
| $C_D$ | f($Re$, ***AR***, $\lambda$, $\Lambda$) | f($Re$, $Ma_\infty$, ***AR***, $\lambda$, $\Lambda$) |
| $C_M$ | f($\alpha$, $\alpha_{\ell 0}$, ***AR***, $\lambda$, $\Lambda$) | |

**Table 2.2** Summary of 2D dependencies for $\alpha < \pm\, 15°$ (***AR*** = ∞).

| | ***Ma* < 0.3** | ***Ma* ≥ 0.3** |
|---|---|---|
| $c_\ell$ | f($\alpha$, $\alpha_{\ell 0}$) | f($Ma_\infty$, $\alpha$, $\alpha_{\ell 0}$) |
| $c_d$ | f($Re$, $\alpha$) | f($Re$, $Ma_\infty$, $\alpha$) |
| $c_m$ | f($\alpha$) | |

**Table 2.3** Summary of pressure coefficient dependencies.

| |
|---|
| 3D $C_L = \oiint (C_{pl} - C_{pu}) d\left(\frac{x}{c}\right) d\left(\frac{y}{b}\right)$ |
| 2D $c_\ell = \oint (C_{pl} - C_{pu}) d\left(\frac{x}{c}\right)$ |
| 3D $C_{M0} = -\oiint (C_{pl} - C_{pu}) \left(\frac{x}{c}\right) d\left(\frac{x}{c}\right) d\left(\frac{y}{b}\right)$ |
| 2D $c_{m0} = -\oint (C_{pl} - C_{pu}) \left(\frac{x}{c}\right) d\left(\frac{x}{c}\right)$ |

For compressible flows, the strictly subsonic or supersonic form of the small perturbation equation in terms of the potential function is given as

$$\boxed{\left(1 - Ma_\infty^2\right)\phi_{xx} + \phi_{yy} = 0} \tag{2.38}$$

## Problems

**2.1** Show that the angle of attack ($\alpha$) and the ratio of specific heats for perfect gases ($\gamma$) are dimensionless by writing out their definition and scrutinizing their units.

**2.2** The maximum value of the lift coefficient (or stall point) is known to depend also on airfoil thickness and in particular on its maximum value ($t_m$). Let us introduce this parameter in our description of the force vector equation (2.12)

$$\vec{F} = \mathrm{f}(\rho_\infty, V_\infty, \mu_\infty, a_\infty, \alpha, c, b, t_m)$$

a) For reference parameters, we need to use a set that contains the three fundamental dimensions with parameters that remain unchanged. Which of the ones shown below are proper?

(i) $V_\infty$, $\mu_\infty$, $c$ (ii) $V_\infty$, $a_\infty$, $b$ (iii) $\mu_\infty$, $\alpha$, $c$

b) Starting with this new form of Eq. (2.12), arrive at the corresponding form of Eq. (2.17) using $\rho_\infty$, $V_\infty$, and $c$ as reference parameters.

**2.3** It was found empirically that in high-speed flows, the effects of compressibility depend strongly on the speed of sound ($a_\infty$) which in turn depends on the pressure ($p_\infty$), the density ($\rho_\infty$), and the ratio of specific heats ($\gamma$). Without using the fact air behaves as a perfect gas, obtain by dimensional reasoning an expression for $a_\infty$ (the dependent parameter). (Note that here we are looking only for dimensionless groupings; we will revisit the actual formulas in Chapter 8.)

**2.4** An un-manned, electric, aerial vehicle originally designed to fly on Earth is being sent to Mars where the atmosphere is very thin and mostly carbon dioxide. The Martian atmospheric density is about 0.020 kg/m$^3$ and the Martian gravity is only 0.38 that of Earth. Assuming that the lift coefficient can remain unchanged since it only depends on $\alpha$ and wing geometry, what parameter would have to change for this vehicle to be able to fly on Mars?

## Check Test

**2.1** Does the physical meaning of potential function ($\phi$) for a fluid differ from the electrical potential function? Would the electrical current then be related to the fluid stream function ($\psi$)?

**2.2** Show that the pressure coefficient, the aspect ratio, and the moment coefficient are dimensionless quantities.

**2.3** A more formal way to introduce the planform area ($S$) into Eq. (2.13) is to multiply $\left(\frac{\vec{F}}{\rho_\infty V_\infty^2 c^2}\right)$ by another dimensionless coefficient. Identify the needed coefficient from those given in Section 2.5.

**2.4** If all angles are dimensionless quantities, what is the difference between *degrees* and *radians*?

**2.5** Arrive at a non-dimensional form for the lift force ($L$) in terms of the free stream pressure ($p_\infty$) and the planform area ($S$).

**2.6** Show that $\phi(x, y) = \frac{1}{2}(x^2 - \beta^2 y^2)$, where $\beta = \sqrt{1 - Ma_\infty^2}$ satisfies the small perturbation equation, Eq. (2.38), for $Ma_\infty < 1.0$.

**2.7** Match the following conditions on the left with *all* corresponding statements on the right (a ➔ k)

| | | |
|---|---|---|
| Incompressible | _____ | a) $gz = 0$ |
| Inviscid | _____ | b) $\nabla \times \vec{V} = 0$ |
| Steady flow | _____ | c) $\mu$ = constant |
| Perfect fluid | _____ | d) d( )/d$t$ = 0 |
| Irrotational | _____ | e) $\tau$ = constant |
| Compressible | _____ | f) $\rho$ = constant |
| Ideal flow | _____ | g) $Ma_\infty \leq 0.3$ |
| | | h) $\Gamma = 0$ |
| | | i) $\mu = 0$ |
| | | j) $Ma_\infty > 0.3$ |
| | | k) no boundary layers |

# 3

# Dynamics of Incompressible Flows

## 3.1 Introduction

Having established the necessary fundamentals, we now examine the dynamic behavior of air as a constant density fluid. This type of flow is the one we most commonly experience and is also a foundation for understanding the more complicated compressible regimes. In this chapter, we will apply the concepts that were detailed in "The most advantageous viscous envelope" in Chapter 1. Working equations for ideal two-dimensional flows are developed here.

A set of flows, called *elementary flows*, are introduced because they are particularly useful in aerodynamic analysis. Since they satisfy Laplace's equation, these flows may be added to generate other more natural flow situations – the classic example being flow around a circular object that represents the section of a cylinder. When rotation is included, such circular cylinder will be shown to develop high lift under ideal flow conditions, but in reality cylinders suffer from high drag that results from flow separation being "totally blunt bodies." In our task to build more complicated flows, the stream function and the potential function, redundant auxiliary functions introduced in Chapter 2 will help visualize the resulting two-dimensional picture.

*Circulation* is a particularly important flow component for the production of lift as mentioned in Chapter 1 and further elaborated in Chapter 4. Without significant amounts of bound circulation, flight would not be as elegant or even possible. The circulation that surrounds the wing results when viscous phenomena in the boundary layers are appropriately distributed over the body's surface and this effect depends on the wing's configuration and angle of attack. We model basic circulation with an elementary flow called the *potential vortex* which is two-dimensional. This elementary flow extends to three dimensions as a *vortex filament* which is useful to model consequences of injecting (shedding) such filaments into flow regions where shearing forces are absent or minimal. The theorems of Helmholtz and Kelvin add valuable concepts to our understanding of the behavior of vortex phenomena in low viscosity fluids such as air. To maintain ideal flow contours, the center of the potential vortex has to be judiciously isolated as discussed in Example 3.3.

## 3.2 Objectives

After successfully completing this chapter, you should be able to:

1) Explain why we focus on inviscid regions outside the boundary layers.
2) Be familiar with the equation for the stream functions of uniform flow of magnitude $V_\infty$, of a doublet of strength $K$, and of a potential vortex of magnitude $\Gamma_0$.

*Elements of Aerodynamics: A Concise Introduction to Physical Concepts,* First Edition. Oscar Biblarz.

Companion website: www.wiley.com/go/elementsofaerodynamics

3) Explain what is meant by "the linearity of Laplace's equation allows the superposition elementary flows that may properly represent the dynamics of an air flow."
4) Using either a velocity potential function or its corresponding stream function demonstrate the ability to find the velocity field in two dimensions (either Cartesian or polar coordinates).
5) Describe in your own words the significance of the mathematical concept of circulation.
6) Find the circulation ($\Gamma$) given a two-dimensional velocity field ($\vec{V}$) for a closed curve encircling the airfoil.
7) Superpose a uniform flow and a doublet and discuss the resulting flow with equations and/or with the aid of diagrams.
8) For each of the three elementary flows in Table 3.1, sketch the corresponding streamlines.
9) Explain how the superposition of the three elementary flows in Table 3.1 may result in flow around a cylinder with circulation.
10) Describe how Helmholtz's and Kelvin's theorems expand the concept of circulation to make it more useful in aerodynamics.

## 3.3 Elementary Flows

Elementary flows are given by mathematically simple equations that capture portions of physical flows of interest. For example, when free vortex flow is superposed onto a uniform flow, a force can develop. Similarly, an object's thickness forms when adding uniform flow to a source and a sink flow of equal strength. The potential vortex when applied over a cylinder results in flow rotation. Viscous effects per se as well as form drag cannot be modelled with elementary flows because these phenomena are not ideal. Circulation, in the real world, is transmitted through the shearing layers at a cylinder's surface so we must very judiciously isolate any flow regions under ideal flow analysis. Similarly, we have to exclude any aft-separation regions from blunt bodies; experiments show that in flows around a cylindrical object the wakes become less pronounced with rotation and on small diameters as wires as discussed in Section 3.7.

In Chapter 2, we introduced the stream function $\psi$ which physically represents lines of fluid flow because streamlines need to be tangent to the velocity vector (automatically satisfying continuity in incompressible flows). The stream function is only defined for two-dimensional flows. When we write a velocity vector as $\vec{V} = u\vec{i} + v\vec{j}$, then in terms of the stream function, the velocity components become

$$u = \frac{\partial \psi}{\partial y} \quad \text{and} \quad v = -\frac{\partial \psi}{\partial x} \tag{2.5}$$

Along a streamline, by definition

$$d\psi = -v dx + u dy = 0 \tag{3.1}$$

We have stated that for irrotational flows (i.e. when $\nabla \times \vec{V} = 0$), $\nabla^2 \psi = 0$ which indicates that both the potential function ($\phi$) and the stream function are governed by Laplace's equation so that any flows described by these functions may be added because of the linearity of their governing equation – this leads to the so-called principle of *superposition of elementary flows* which is most useful for building more complicated flows that are inviscid, incompressible, and irrotational. Moreover, $\psi$-lines and $\varphi$-lines should be easily visualized because they form "perpendicular grids" as depicted in Figure 3.1.

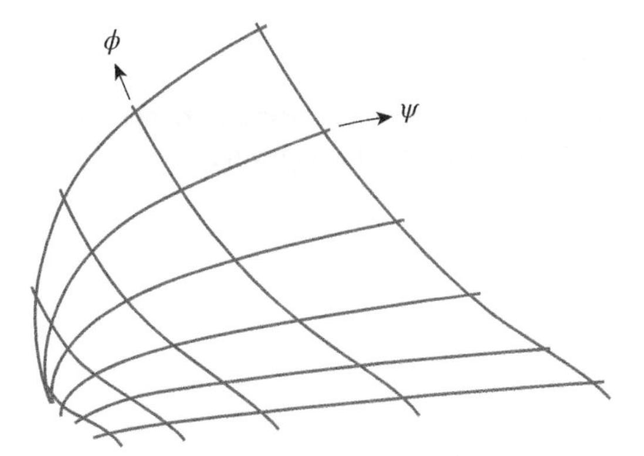


Figure 3.1 The potential and stream functions intersect perpendicular to each other forming a rectangular grid or more generally curvilinear squares.

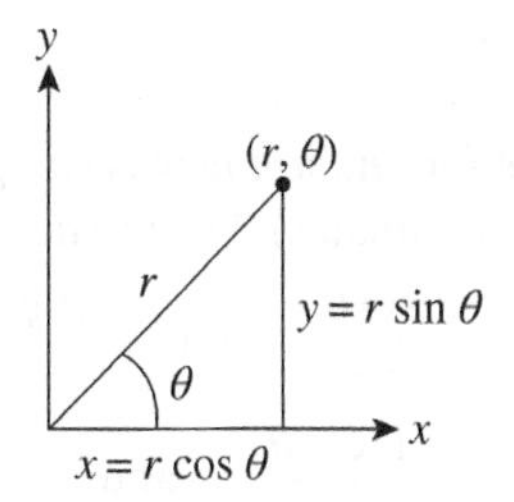


Figure 3.2 Cartesian and polar coordinate equivalents used in two-dimensional elementary flow descriptions.

In aerodynamics, the most useful two-dimensional elementary flows are:

i) Uniform flow
ii) Vortex flow
iii) Sources/sinks (doublets)

We will incorporate uniform flow and vortex flow within the so-called *thin airfoil theory* of Chapters 5 and 6 as it requires vanishingly thin surfaces. In this chapter, however, we need the third elementary flow above in its doublet form (where sources and sinks are of equal strength).

In order to study the flow around a cylindrical section and other such two-dimensional objects, we find it more convenient to work with our elementary flows in polar coordinates $(r, \theta)$. Figure 3.2 shows coordinate equivalents between Cartesian and polar.

Table 3.1 gives the appropriate set of equations for uniform flow, the doublet, and the potential vortex. Both potential and the stream functions are identified for each.

Table 3.1 Elementary flows in polar coordinates.

| Type of flow | Stream function $\psi$ | | Potential function $\varphi$ | |
|---|---|---|---|---|
| Uniform<br>$u = V_\infty,\ v = 0$ | $V_\infty r \sin\theta$ | (3.2) | $V_\infty r \cos\theta$ | (3.3) |
| Doublet<br>$K =$ doublet strength | $-\dfrac{K \sin\theta}{2\pi r}$ | (3.4) | $\dfrac{K \cos\theta}{2\pi r}$ | (3.5) |
| Clockwise potential vortex<br>$V_r = 0,\ \ V_\theta = -\dfrac{\Gamma_0}{2\pi r}$<br>$\Gamma_0 =$ Circulation strength | $\dfrac{\Gamma_0}{2\pi}\ln r$ | (3.6) | $-\dfrac{\Gamma_0 \theta}{2\pi r}$ | (3.7) |

**Example 3.1** Show that each of the three flows given in Table 3.1 satisfies Laplace's differential equation.

We will work with the stream function here but the same steps apply to the potential function. The two-dimensional Laplace's equation is given in both coordinate systems of interest as

$$\frac{\partial^2 \psi}{\partial x^2} + \frac{\partial^2 \psi}{\partial y^2} = 0 \qquad \text{Cartesian coordinates}$$

$$\frac{\partial^2 \psi}{\partial r^2} + \frac{1}{r}\frac{\partial \psi}{\partial r} + \frac{1}{r^2}\frac{\partial^2 \psi}{\partial \theta^2} = 0 \qquad \text{Polar coordinates}$$

The proof for *uniform flow* is straightforward and will be left to the reader (Hint: work here with Cartesian coordinates). For the *doublet* in polar coordinates, we have the following:

$$\frac{\partial \psi}{\partial r} = \frac{K \sin\theta}{2\pi r^2} \quad \text{and} \quad \frac{\partial^2 \psi}{\partial r^2} = -\frac{K \sin\theta}{2\pi r^3}$$

$$\frac{\partial \psi}{\partial \theta} = -\frac{K \cos\theta}{2\pi r} \quad \text{and} \quad \frac{\partial^2 \psi}{\partial \theta^2} = \frac{K \sin\theta}{2\pi r}$$

$$\left(-\frac{K \sin\theta}{\pi r^3} + \frac{1}{r}\frac{K \sin\theta}{2\pi r^2}\right) + \left(\frac{K \sin\theta}{2\pi r^3}\right) = 0$$

Now for the *potential vortex* (notice that there is no $\theta$-dependence)

$$\frac{\partial \psi}{\partial r} = \frac{\Gamma_0}{2\pi r} \quad \text{and} \quad \frac{\partial^2 \psi}{\partial r^2} = -\frac{\Gamma_0}{2\pi r^2}$$

$$-\frac{\Gamma_0}{2\pi r^2} + \frac{1}{r}\frac{\Gamma_0}{2\pi r} = 0$$

This confirms that these elementary flows satisfy Laplace's equation. When adding (or superposing) them, the resultant flows also satisfy Laplace's equation.

*Uniform flow* is the simplest but most important component in Table 3.1 because it represents the incoming air stream relative velocity which is indispensable in our work. Unless stated otherwise, $V_\infty$ is always in the positive $x$-direction and time independent.

The *doublet* results from the sum of a point-fluid-source with a point-fluid-sink of equal strength, the two being brought to a common origin. As the distance between them goes to zero, their strengths must grow so as to keep the product of the two strengths finite, as depicted in the streamlines on the right side of in Figure 3.3. In this figure and as shown in Example 3.2, such doublet arrangement conveniently generates a cylindrical cross-section when uniform flow is added, and this configuration is convenient for analyzing ideal flow around a such object. More generally, in a uniform flow with sources and sinks that remain separate, we can generate oval shapes as long as they both are of equal strength.

The *potential vortex* is another flow that is best described in polar coordinates. We can envision that when a clockwise potential vortex is added to uniform flow coming from the left, the net effect is to

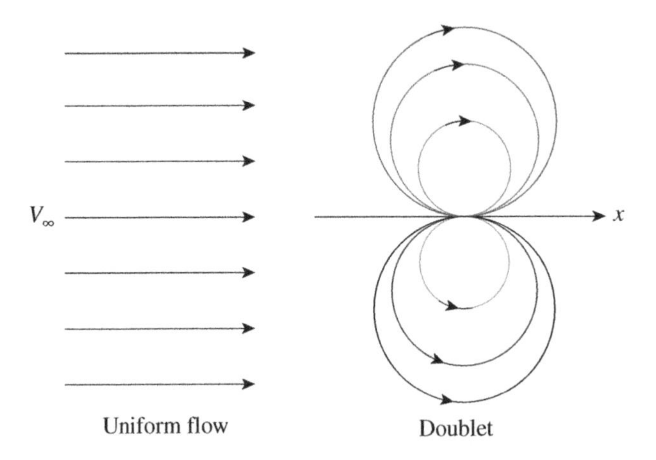


**Figure 3.3** Streamlines of uniform flow approaching a doublet along the $x$-axis. As shown, the doublet streamlines issue from a point toward left and merge toward the right so as to generate a closed surface. The source and sink flows must be of equal strength.

speed up the flow above and slow down it down below with corresponding effects on the pressures. The minus sign is needed so that when clockwise circulation is added with uniform flow moving to the right, the resultant pressure force from Bernoulli's equation is a vertical force or lift. As might be apparent from Table 3.1, the ideal potential vortex has infinite vorticity at $r = 0$. With time, in the real world any such high value dissipates by the diffusion of vorticity from the core. In nature, large vortex flows are seen in tornados and hurricanes and small vortex flows or eddies in the turbulent flows. For airfoil flows, the source of vorticity is located at the boundary layers with another source at the wing tips. Other relevant details for the potential vortex are given in Section 3.4.

**Example 3.2** Construct the flow around a circular cylinder by adding a *doublet* to a *uniform flow* as in Figure 3.3.

This can be done using two streamline functions from Table 3.1.

$$\psi = V_\infty r \sin\theta - \frac{K \sin\theta}{2\pi r} = \sin\theta\left(V_\infty r - \frac{K}{2\pi r}\right)$$

We solve for $\psi = 0$ to represent the constant streamline that bounds the object. Equating the above sum to zero makes either $\sin\theta = 0$ or $\left(V_\infty r - \frac{K}{2\pi r}\right) = 0$. The former means that $\theta = 0$ or $2\pi$ (the stagnation points) and the later gives a cylinder radius $r \equiv R = \sqrt{\frac{K}{2\pi V_\infty}}$. This technique can be used to make other similar closed shapes. As already mentioned, an oval shape consists of a source separated from a sink of equal strength both immersed in a uniform flow.

## 3.4 Circulation

We are now ready to formalize the concept of the *circulation* which is given the symbol ($\Gamma$) and as the name implies is the feat of encircling some region in space as seen in Figure 3.4.

Mathematically, *clockwise circulation* is considered negative by definition and given by

$$\Gamma \equiv -\oint_C \vec{V} \cdot d\vec{r} = -\iint_S \left(\nabla \times \vec{V}\right) \cdot d\vec{A} \tag{3.8}$$

The line integral around a closed contour "$C$" in the first integral is shown to be equivalent to a surface integral of the curl of $\vec{V}$ over the enclosed surface "$S$," a form which will be more useful in some cases. The second integral is arrived at by applying *Green's theorem* in the plane and proof may be found in standard calculus references. Equation (3.8) shows how *circulation* and *vorticity* are mathematically connected.

Relations in (3.8) may be written in two-dimensional Cartesian form using their corresponding vector components $\vec{r} = [x, y]$ and $\vec{V} = [u, v]$

$$\Gamma = \oint (u dx + v dy) \tag{3.9}$$

$$\nabla \times \vec{V} = (\partial v/\partial x - \partial u/\partial y)\hat{k} \tag{3.10}$$

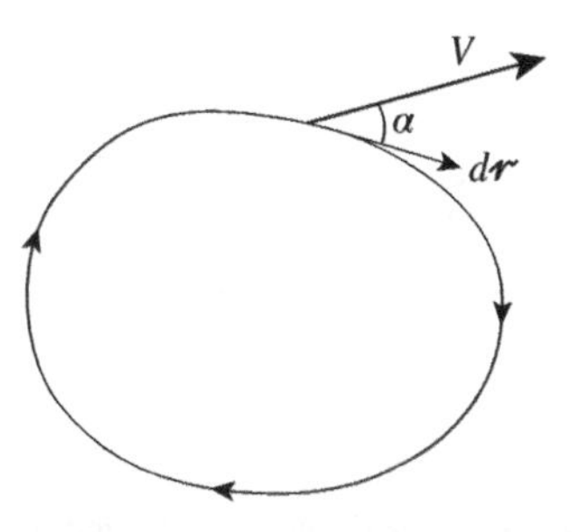

**Figure 3.4** Clockwise circulation contour used for the generation of lift.

We continue by examining some well-known rotational flow examples. Suppose that

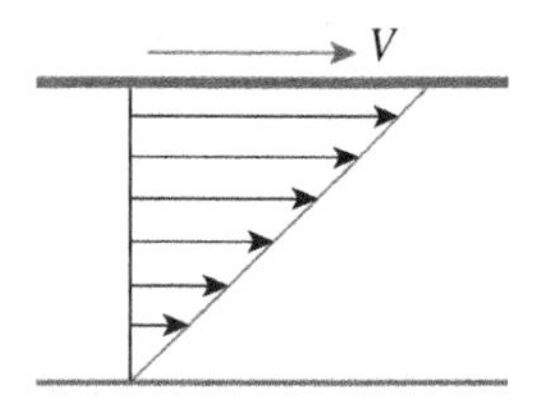


Figure 3.5 Couette flow between flat plates.

$$v = \kappa_1 x \quad \text{and} \quad u = \kappa_2 y \tag{3.11}$$

where $\kappa_1$ and $\kappa_2$ are constants. When $\kappa_1 \neq \kappa_2$ this set represents a variety of *shearing motions*. In particular, when $\kappa_1 = 0$, we obtain one-dimensional flow where $u$ is a linear function of $y$ which is the well-known *plane Couette flow,* i.e., flow between a stationary lower plate and a moving upper parallel plate which through the action of the intermediate's fluid viscosity imparts movement between parallel fluid layers (more of this in Chapter 7). See Figure 3.5.

Furthermore, when $-\kappa_1 = \kappa_2 \equiv \omega$, we have$\nabla \times \vec{V} = \omega \vec{k}$ or $V_\theta = \omega r$ which is the equation for *solid-body rotation* (depicted in Figure 3.6) – solid-body rotation is included here because it should be familiar from solid-body dynamics. Last but not least, the *potential vortex* is given by a rotational velocity that is inversely proportional to its polar-coordinate radius $r$ as seen in Table 3.1 where it is written as $V_\theta = -\dfrac{\Gamma_0}{2\pi r}$. Rotational velocity ($V_\theta$) profiles for both solid-body rotation and the potential vortex are compared in Figure 3.6 as a function of the radius $r$.

Vorticity in a *potential vortex* resides only at its core. This can be shown by calculating the circulation (defined in Eq. (3.8)) around a closed path that encloses the vortex origin. Any closed path that excludes this origin, however, can be shown to have no circulation, and this is a useful ploy for applying our ideal equations in flows which contain such rotational components. In order to avoid vortex origins, we also need to draw carefully crafted contours to exclude the thin boundary layers (see Figure 2.5) together with their wake. In succeeding chapters, we will expand the concept of the two-dimensional potential vortex to one or a series of *vortex filaments*

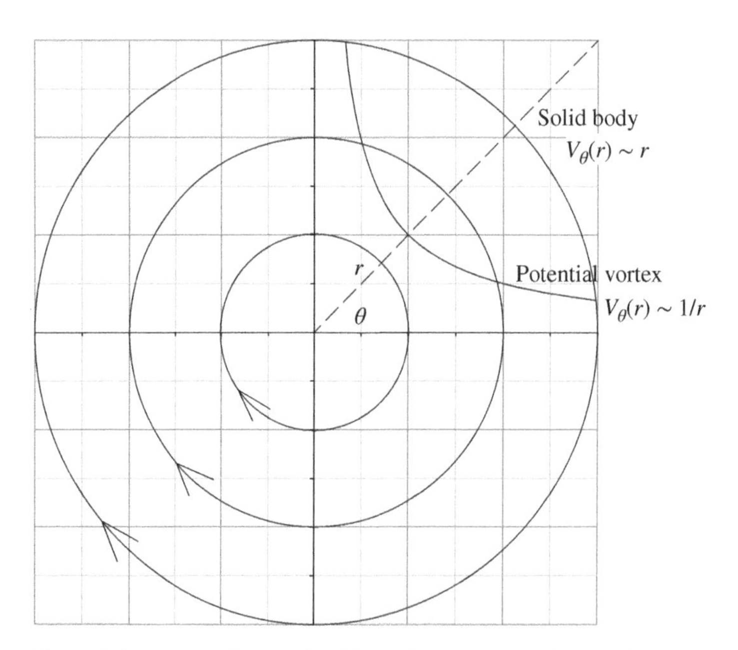


Figure 3.6 A *vortex filament* is different from a more ordinary *solid-body rotation* in which $V_r = 0$ and $V_\theta = \omega r$. A turntable and a spinning top are examples of solid-body rotation (dashed line), whereas an atmospheric tornado is an example of a potential vortex (solid line).

to represent the physical region that coincides with the axis of rotation of successive fluid elements. As shown in Chapter 5, a number of tiny vortex filaments distributed over an airfoil's camber line when added to a uniform free stream flow can model a lifting-line element. We reiterate that our license to applying Bernoulli's equation is based on using the pressures at the boundary layer's edge with the viscous region itself being carefully excluded (though present to configure the Kutta condition).

**Example 3.3** Using Eq. (3.8) show that vortex flows have a net circulation only when we include their origin.

From Table 3.1, in polar coordinates the potential vortex has no radial velocity $V_r$ and its circumferential velocity component is $V_\theta = -\dfrac{\Gamma_0}{2\pi r}$. If we take a clockwise circular contour at some radius $r$ hinged at the vortex origin, Eq. (3.8) becomes

$$\Gamma \equiv -\oint_C \vec{V}\cdot d\vec{r} = -\oint\left(\frac{\Gamma_0}{2\pi r}\right) r d\theta = \frac{\Gamma_0}{2\pi}\int_0^{2\pi} d\theta \equiv \Gamma_0$$

Notice the radius cancelled out so that the result is the same for any size radius circular contour as long as it encloses the vortex origin. Here the reason for the choosing this flow's constant as $\Gamma_0/2\pi$ becomes apparent.

The choice of polar coordinates makes the above integration straightforward. For contours that do not include the origin, we need to resort to integrating along a pair of circles that include a small "cut" excluding the origin and these need to double back to close the contour. Such a path would sweep the origin back and forth and the resulting integral would be zero further demonstrating that all the circulation in this flow resides at the origin.

## 3.5 Superposition of Elementary Flows

We have already shown in Example 3.2 that the addition of uniform flow and doublet flow results in fluid motion around an object shaped as a circular cylinder. Under ideal conditions, however, no net force should develop because of the total symmetry of the resulting pressure distributions around the cylinder. A more consequential case involves the addition of a potential vortex at the axis of a doublet in combination with the uniform flow. The reason for this is that the vortex adds circulation to the flow surrounding the circular cylinder thus distorting the pressure symmetry. Let us show some of these details by adding the three stream functions in Table 3.1:

$$\psi_{TOT} = V_\infty r\sin\theta - \frac{K\sin\theta}{2\pi r} - \frac{\Gamma_0}{2\pi}\ln r \tag{3.12}$$

$$V_r = V_\infty \cos\theta\left(1 - \frac{K}{2\pi V_\infty r^2}\right) \tag{3.13}$$

$$V_\theta = \frac{1}{r}\left(-V_\infty r\sin\theta - \frac{K\sin\theta}{2\pi r} - \frac{\Gamma_0}{2\pi}\right) \tag{3.14}$$

In Eq. (3.13), setting $V_r = 0$ fixes the shape of the immersed object because there can be no flow across its surface and it turns out to be the same cylinder shown in Example 3.2 before circulation was added. The cylinder radius follows from Eq. (3.13) as $r_s \equiv R = \sqrt{K/2\pi V_\infty}$ and Eqs. (3.15) and (3.16) can now be written in terms of this cylinder radius $R$.

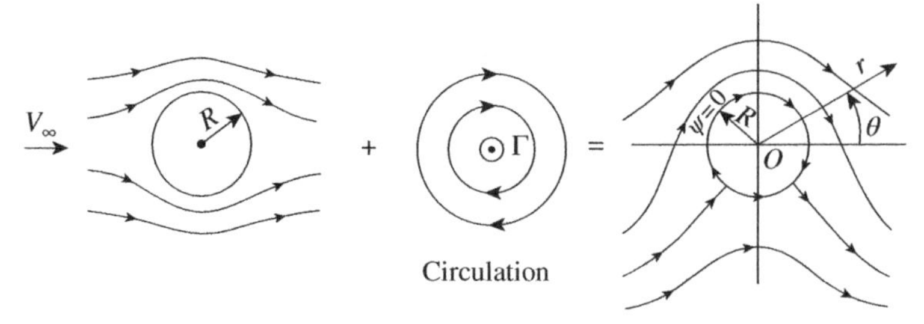


**Figure 3.7** Superposition of uniform flow over a doublet with circulation.

$$V_r = V_\infty \cos\theta \left(1 - \frac{R^2}{r^2}\right) \tag{3.15}$$

$$V_\theta = -V_\infty \sin\theta \left(1 + \frac{R^2}{r^2}\right) - \frac{\Gamma_0}{2\pi r} \tag{3.16}$$

According to Eq. (3.14) without circulation, $V_\theta$ would vanish when $\sin\theta_S = 0$, namely, at both $\theta_S = 0°$ and $180°$, which are the same stagnation points on the cylinder surface found in Example 3.2. However, the stagnation point locations will change with increasing $\Gamma_0$ and move down as shown in Figure 3.7. With $\Gamma_0 = 4\pi R V_\infty$, the stagnation points merge at $\theta_S = 270°$; increasing $\Gamma_0$ further causes the flow to leave the surface so $\theta_S = 270°$ is the stagnation point for maximum circulation on the cylinder. Such distorted pressure rearrangements generate a lift force but no drag due to the *vertical symmetry* of the resulting streamlines as noted in the figure. Further details on how the sectional lift force emerges are given in Chapter 4 where we actually calculate lift. When dealing with ideal flows, there can be no wake from the cylinder and complete symmetry is retained about its vertical axis even under circulation.

## 3.6 Theorems of Helmholtz and Kelvin

The potential vortex has other important applications in aerodynamics. As stated earlier, to extend this two-dimensional elementary flow to three dimensions, we introduce the concept of a *vortex filament*. Such filaments incorporate the line connecting all centers of two-dimensional potential vortices which coincide with successive fluid elements. A helpful feature of the vortex filament is that may curve in space. With the help of the Biot–Savart law of electromagnetics, we can calculate a velocity induced by all elements of the vortex filament at any chosen location and show that the tangential velocity $V_\theta$ in Table 3.1 is representative of a slice along a vortex filament.

Helmholtz's theorems state relevant properties of vortex filaments such as: any circulation ($\Gamma_0$) present must be constant along the vortex filament length and that such filaments cannot start or end in an inviscid fluid but must form a closed path, akin to a smoke ring, or end at a solid boundary. Also, in ideal fluids, no fluid particle can have rotation if it did not originally rotate. In real fluids like air and water, vorticity is shed from regions of fluid rotation into the non-shearing portions of the flow where these vortices are preserved for sufficiently long times to affect pressure distributions around nearby objects.

Kelvin's theorem states that within ideal flow regions surrounding an airfoil, the time derivative of the circulation must vanish, or $d\Gamma/dt = 0$. The proof involves taking the time derivative of Eq. (3.8)

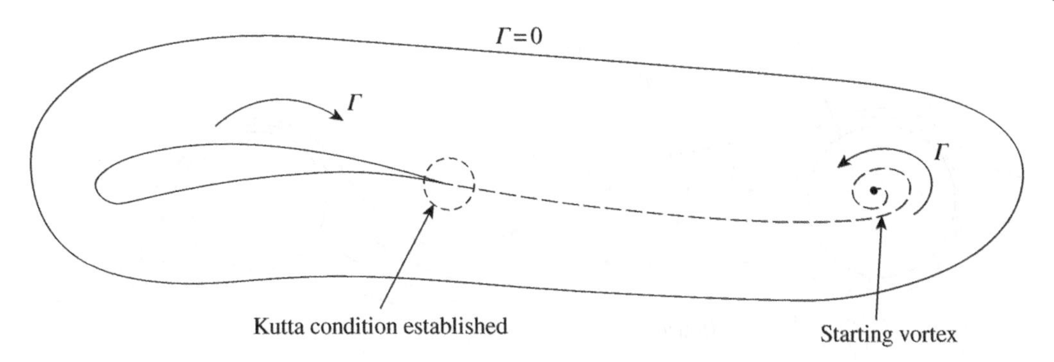


**Figure 3.8** Shedding of a starting vortex that according to Kelvin's theorem makes the total circulation in ideal fluids to be zero inside the drawn contour.

and requires following a given fluid element as a *control mass* entity. We use a special form of the momentum equation along a streamline for $d\vec{V}/dt$ (shown later in Chapter 4 as Eq. (4.2)). Briefly,

$$-\frac{d\Gamma}{dt} = \frac{d}{dt}\oint \vec{V} \cdot d\vec{r} = \oint\left[\frac{d\vec{V}}{dt} \cdot d\vec{r} + \vec{V} \cdot d\vec{V}\right] \tag{3.17}$$

$$-\frac{d\Gamma}{dt} = \oint\left[-\frac{\nabla p}{\rho} \cdot d\vec{r} + d\left(\frac{V^2}{2}\right)\right] \equiv 0 \tag{3.18}$$

Both integrals on the right-hand side of Eq. (3.18) go to zero, as they are point functions integrated along a closed contour (remember that the density is a constant). The circulation, therefore, remains constant inside any such ideal fluid contour. The product ($\nabla p \cdot d\vec{r}$) may also be written as $dp$ to recognize it as a point function. Note that we are here using the momentum equation without any shearing forces as if the fluid is inviscid and that $d\vec{r}/dt = d\vec{V}$.

To sum up, Kelvin's theorem implies that for ideal fluids to remain irrotational any rotation issuing from an upstream boundary layer must continue its path downstream unchanged. This leads to the concept of the *starting vortex* shown in Figure 3.8 which is shed into the airstream *whenever an airfoil accelerates or is impulsively started* – any *sudden* generation of lift produces a large starting vortex that remains in the air in the same location. Shed stating vortices last long enough to require careful aircraft spacing on the runway during the takeoff of jumbo jets and other large aircraft because even during non-impulsive takeoffs starting vortices are continuously being generated. In time and in the absence of other vortex sources, the small air viscosity does dissipate this shed circulation.

## 3.7 Real Flows

Flows around ordinary blunt bodies unavoidably separate at their wake and the cylinder is no exception. The simplest explanation for this is that inside the boundary layers some flow energy is continuously being dissipated along the flow path. At the forward portion of the cylinder, the pressure is decreasing as the flow accelerates but after the crest the flow decelerates resulting in an *adverse pressure gradient* being developed. Because the flow has lost a portion of the energy needed to climb this "pressure hill," it tends to easily separate. Under viscous conditions, there are differences between high and low Reynolds number flows as well between a laminar and

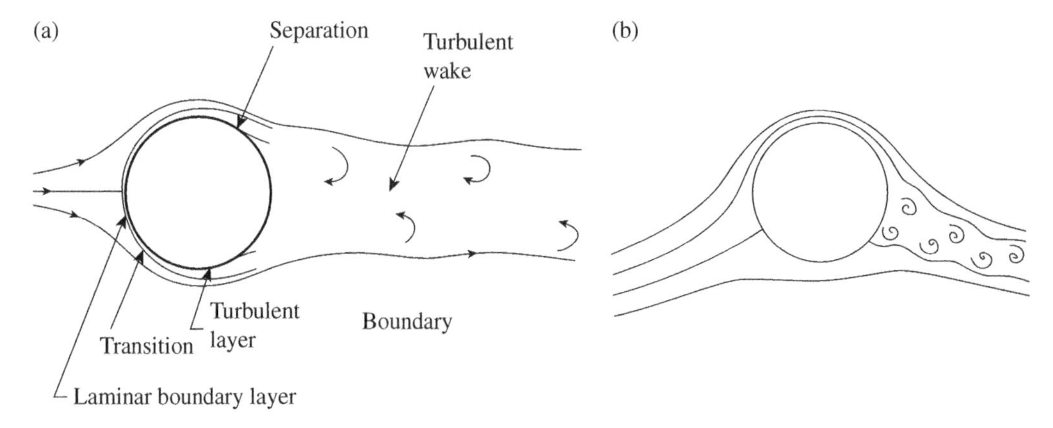


**Figure 3.9** Real flows around a circular cylinder. Notice the effect of rotation on diminishing the separated region. (a) Non-rotating. (b) Rotating.

turbulence flows – these topics are treated in Chapter 7. In real fluids, a rotating cylinder will transmit its rotation through its adjacent shear layers and, as shown in Figure 3.9, flow separation at the cylinder's wake is reduced by this fluid rotation. A separation decrease is more readily accomplished by reducing the rotating cylinder's diameter to the size of a very thin wire.

## 3.8 Summary

The topics covered in this chapter bring us one step closer to our goal of analytically formulating the lift on an airfoil. The concept of elementary flows and the results of their superposition play a vital role in developing the relevant formulations. We have conjectured that one effect of added vorticity is to modify ideal-flow streamlines so that resulting pressure distributions generate a lift force and this remains to be shown. The role of the potential vortex filament introduced in this chapter is expanded in Chapter 4, and in Chapter 5 vortex filaments are used as a basic component in thin airfoil theory. By analyzing flow models with thin rotating wires, we can better represent ideal conditions where wake-flow separation tends to be minimal.

Circulation is an important concept and is mathematically given by the integral of the velocity vector around a suitable closed path. It is also related to $\nabla \times \vec{V}$ as

$$\Gamma \equiv -\oint_C \vec{V} \cdot d\vec{r} = -\iint_S \left(\nabla \times \vec{V}\right) \cdot d\vec{\mathbb{A}} \tag{3.8}$$

Inside contours where the vorticity is zero we have irrotational flows so in order to use irrotational formulas like Bernoulli's equation we need to carefully construct the regions with Eq. (3.8).

The sum of uniform flow with a doublet and a clockwise vortex located at the same origin generates ideal flow streamlines around a cylindrical object of radius $R$ whose polar velocity components are

$$V_r = V_\infty \cos\theta \left(1 - \frac{R^2}{r^2}\right) \tag{3.15}$$

$$V_\theta = -V_\infty \sin\theta \left(1 + \frac{R^2}{r^2}\right) - \frac{\Gamma_0}{2\pi r} \tag{3.16}$$

Note that these equations represent ideal flow around a cylinder for $r \geq R$.

## Problems

For all these problems, assume that the flow is two-dimensional and incompressible.

**3.1** For the flow $u = 2x + y$ and $v = -2y + x$, find the potential and stream functions. What can we say about this flow satisfying continuity and being irrotational?

**3.2** A flow field is described by $V(x, y) = \omega\sqrt{x^2 + y^2}$ with streamlines $x^2 + y^2 = c^2$. Both $\omega$ and $c$ are constants. Is this flow irrotational?

**3.3** Given the stream function

$$\psi = \theta + \ln(r)$$

a) Does this flow satisfy continuity when the fluid is incompressible?
b) Is this flow rotational?

**3.4** Superpose a uniform flow in the positive $x$-direction of constant magnitude $U_0$ with a uniform flow in the positive $y$-direction of magnitude $V_0$. Give the resulting stream and potential functions and sketch the flow.

**3.5** Superpose the flow represented below and describe the resulting flow in terms of its streamlines. Sketch some streamlines in the $x$–$y$ plane.

$$\phi_1 = \frac{1}{2}x^2$$

$$\phi_2 = -\frac{1}{2}y^2$$

## Check Test

**3.1** Give two examples of common constant density flows in air.

**3.2** Write the Cartesian coordinate form of the vector operation $\nabla \times \vec{V}$ in two-dimensional flows.

**3.3** Why do a source and a sink of equal strengths when added to a free stream flow produce a closed two-dimensional object?

**3.4** When a doublet is added to uniform flow, we generate a cylindrical object. Does the resulting flow around it satisfy continuity?

**3.5** Why must every portion of a flow considered to be "ideal" be irrotational?

**3.6** Can contours of equal elevation around a mountain be thought as potential functions?

**3.7** Can lines of current across a discharging capacitor be considered streamlines?

# 4

# Mass, Momentum, and Energy Principles

## 4.1 Introduction

In previous chapters, we introduced principles that govern aerodynamic phenomena, specified necessary airfoil attributes, and developed the most common form of the dimensionless coefficients in aerodynamics. We are now ready to establish using Bernoulli's equation the relationship between flow velocities and their accompanying pressure fields in incompressible flows. This will be followed by proof of the most important theoretical result associated with aerodynamic lift – the Kutta–Joukowski theorem. The fact Bernoulli's equation applies along streamline flows, preferably in locations where the flow is irrotational, requires some clever profiling of the singularities introduced by vortex filaments. Application of Bernoulli's equation and the electromagnetic analog of the Kutta–Joukowski theorem are presented here.

## 4.2 Objectives

After successfully completing this chapter, you should be able to:

1) Explain equation (4.3) in your own words and list all assumptions under which it is valid.
2) Explain why we can treat Bernoulli's equation along a streamline as singly-dimensional.
3) For the airspeed indicator, demonstrate the ability to calculate the (true) airspeed given the output of a Pitot-static instrument and the local density.
4) Write the pressure coefficient for incompressible, irrotational, steady flow when body forces are negligible.
5) Draw pressure coefficient distributions around a circular cylinder and contrast ideal flow with real flow features.
6) Explain why the pressure distributions around a circular cylinder arising from ideal flows without circulation generate neither lift nor drag.
7) Draw pressure coefficient profiles around a cylinder under ideal flow with circulation and identify features that lead to the generation of lift.
8) State in your own words the Kutta–Joukowski theorem highlighting its importance.

## 4.3 Bernoulli's Equation

We will specialize the momentum or Euler equation given in Chapter 2 and shown below. Keep in mind that viscous and other force terms are missing from such formulation of the momentum equation because we are examining flows close to but outside $t$ boundary layers.

*Elements of Aerodynamics: A Concise Introduction to Physical Concepts*, First Edition. Oscar Biblarz.

Companion website: www.wiley.com/go/elementsofaerodynamics

$$\rho \frac{\partial \vec{V}}{\partial t} + \rho\left(\vec{V} \cdot \nabla\right)\vec{V} = -\nabla p \tag{2.10}$$

We proceed by focusing on steady, irrotational flows and applying additional del-vector identities from vector calculus. The convective portion on the left-hand side of this equation may be written as

$$\left(\vec{V} \cdot \nabla\right)\vec{V} = \nabla\left(\frac{V^2}{2}\right) - \vec{V} \times \left(\nabla \times \vec{V}\right) \tag{4.1}$$

The last term in Eq. (4.1) becomes zero when the flow is irrotational so that for *steady flow*, Eq. (2.10) may now be written along the flow direction, i.e. in a direction following the streamline coordinate $r(x, y)$ which may be straight or curved, as

$$\begin{gathered} \rho\nabla\left(\frac{V^2}{2}\right) = -\nabla p \quad \text{or} \quad \rho\frac{d}{dr}\left(\frac{V^2}{2}\right) = -\frac{dp}{dr} \\ d\left(\frac{V^2}{2}\right) + \frac{dp}{\rho} = 0 \end{gathered} \tag{4.2}$$

Integrating Eq. (4.2) between two arbitrary flow locations keeping the density constant and using $p_t$ as a constant representing the *stagnation pressure* or *zero-velocity* condition we obtain

$$\left(\frac{V_1^2}{2}\right) + \frac{p_1}{\rho} = \left(\frac{V_2^2}{2}\right) + \frac{p_2}{\rho} \equiv \frac{p_t}{\rho} \tag{4.3}$$

It should be noted that the stagnation pressure in Eq. (4.3) is constant along a streamline but may change between streamlines in flows that have measurable amounts of rotation. In the given form, Bernoulli's equation is also applicable to liquids but for them a "potential energy term" ($\rho g z$ where $g$ is the local gravity constant and $z$ the elevation above or below some reference location) is added to the sum because liquids like water are 1000-times heavier than air. With this body force term included, we obtain, after multiplying out the density, the most standard form of Bernoulli's equation between two arbitrary locations along a streamline,

$$\frac{1}{2}\rho V_1^2 + p_1 + \rho g z_1 = \frac{1}{2}\rho V_2^2 + p_2 + \rho g z_2 \equiv p_t \tag{4.4}$$

Returning now to gas flows with $Ma < 0.3$, the *pressure coefficient* defined in Chapter 2 (Eq. (2.29)) can now be written for incompressible flows in terms of flow velocities. In Chapter 2, we have already introduced the convective term in Eq. (4.4) and at the free stream given it the symbol $q_\infty = \frac{1}{2}\rho_\infty V_\infty^2$, calling it the *dynamic pressure*. Neglecting elevation changes in Bernoulli's equation, we arrive at the incompressible form of the pressure coefficient with respect to free stream conditions as

$$C_p \equiv \frac{p - p_\infty}{q_\infty} = \frac{\frac{1}{2}\rho_\infty\left(V_\infty^2 - V^2\right)}{\frac{1}{2}\rho_\infty V_\infty^2} = 1 - \left(\frac{V}{V_\infty}\right)^2 \tag{4.5}$$

A version similar to Eq. (4.4) that comes from mechanical-energy considerations is given Section 4.6. Because incompressible flows are very common in aerodynamics (always relevant during takeoff and landing), it is important to know how to use Bernoulli's equation and to understand its applications – one of which is found in airspeed measurement devices.

## 4.4 Airspeed Indicator

The form of Bernoulli's equation given as Eq. (4.3) is used for interpreting measurements from the *airspeed indicator* shown in Figure 4.1. This device is based on a "Pitot-static system" which simultaneously measures static ($p_\infty$) and stagnation ($p_t$) pressures at a fixed location when properly oriented to face an air flow. Rewriting Eq. (4.3) keeping the subscript $t$ for the stagnation condition but denoting the free stream with the subscript $\infty$, we get Eq. (4.6)

$$V_\infty = \sqrt{\frac{2(p_t - p_\infty)}{\rho_\infty}} \tag{4.6}$$

On many aircraft, static pressure measurements ($p_\infty$) more typically take place at a separate location since it can be shown that the free stream pressure is essentially impressed across any thin boundary layers; but, in Figure 4.1, we show a unit that measures both pressures where it is immersed in the flow. In Pitot-static instruments, however, it is important that the static and stagnation ports be close to each other for best accuracy. Example 4.1 elaborates further on the use of the instrument shown in Figure 4.1.

As shown, the instrument in Figure 4.1 is connected to a liquid manometer that outputs $h_m$ as the height difference which can be related to the pressure difference with the fluid statics relation. There are several inherent errors in Pitot-static systems beyond typical instrumentation and positioning errors. The most significant is based on the fact that pressure sensors are normally calibrated at or near sea-level conditions, whereas aircraft usually operates at high enough altitudes for the local density to be different from sea level – at altitude, Pitot-static readings are called true airspeed (*TAS*) in contrast to equivalent airspeed (*EAS*) which is why the instrument is calibrated to read. The measurement correction is found from

$$TAS = EAS\sqrt{\frac{\rho_{sl}}{\rho_\infty}} \tag{4.7}$$

where the density $\rho_{sl}$=at sea level and $\rho_\infty$ = at altitude

The above calculation requires knowledge of air density with elevation, information that can be found in from *density-altitude charts* or equivalent data (see Figure 1.5 and/or Appendix A). Values of the *TAS* are generally about 10–15% larger than EAS.

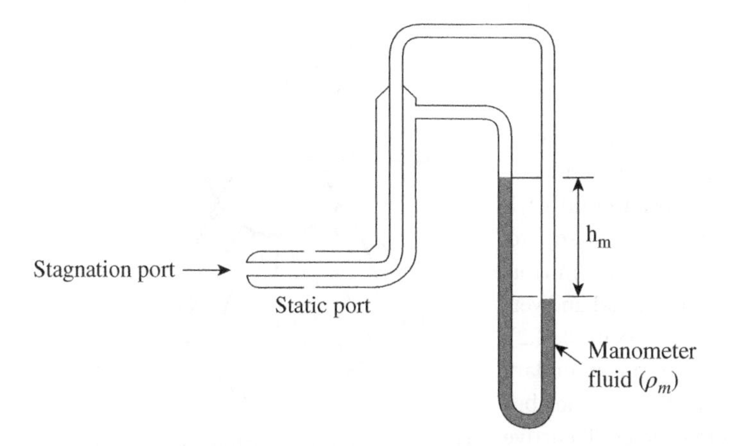


**Figure 4.1** Pitot-static system used for airspeed measurements. The instrument output is $h_m$.

**Example 4.1** An airspeed indicator unit is mounted for calibration in an old wind tunnel where pressure readouts can be given by two U-tube manometers, one using mercury and the other one with pure water. If air at standard-conditions flows in the wind tunnel at 180 km/h (50 m/s), what height $h_m$ will each of the two manometers read?

Using the form of Eq. (4.4), we apply it to the liquid manometers to arrive at Eq. (E4.1) below where both pressures are reflected in elevation or the liquid height, $h_m \equiv z_2 - z_1$. This results in

$$h_m = \left(\frac{V_\infty^2}{2g}\right)\frac{\rho_\infty}{\rho_m} \tag{E4.1}$$

Here g is the local acceleration of gravity and $\rho_m$ the density of the liquid in the U-tube manometer. In liquid density tables, we find that $\rho_m = 13.633 \times 10^3$ kg/m$^3$ for mercury and $1.0 \times 10^3$ kg/m$^3$ for water, and the air density $\rho_\infty$ is taken at standard conditions. Thus

$$h_m = \left(\frac{50^2}{2 \times 9.81}\right)\frac{1.226}{\rho_m} = 1.145 \text{ cm in mercury and } 15.622 \text{ cm in water}$$

For the given velocity and gas conditions, the water manometer would suffice but the mercury manometer would be more appropriate for much higher flow speeds. Modern wind tunnels use solid-state pressure transducers with various calibrated ranges of accuracy and feature built-in electrical outputs usually connected to digital computers for readout/analysis and recording.

## 4.5 Kutta–Joukowski Theorem

In Chapter 1, we introduced the Kutta condition as a necessary ingredient for airfoil lift and called it one of our default assumptions [what were the others?] and the fundamental theorem we are about to introduce is related to it. In order to capture the essence of how lift is generated in incompressible flows, we need to show that the necessary *elementary flows* for the theoretical description of lift consist of a *vortex filament flow* immersed perpendicular to a *uniform flow* (the *doublet* is needed to generate the cylindrical object as shown in Example 3.2). We do this proof using two-dimensional, polar representations for flows around the cylinder, and in Chapter 5, we will expand these concepts to more representative configurations on airfoil sections. Carefully scrutinizing Figure 4.2, we see that any ideal flow around a cylinder without circulation does not generate any *net forces* because pressure distributions are "mirror-like" or symmetric about both the horizontal and vertical axes and thus forces cancel out (see Example 4.2). In real flows, it is well known that the boundary layers separate at the aft-region of the cylinder but such separation only result in drag forces. To arrive at the lift force, we need to continue by only focusing on ideal features of the flow.

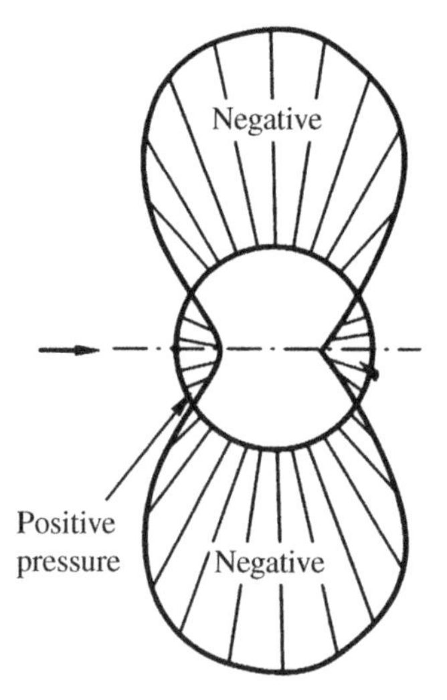


**Figure 4.2** Pressure distributions around a cylinder without circulation. Positive pressures extend into the circle and negative ones out.

To demonstrate that the three *elementary flows* introduced in Chapter 3 are sufficient to extract the sectional lift force, we proceed with examining a cylindrical surface of radius $R$ as described in Chapter 3. Equations (4.8) and (4.9) show the resulting velocity components $V_r$ and $V_\theta$ at the *surface* of the cylinder with circulation (recall in polar coordinates $x = r \cos \theta$ and $y = r\sin \theta$, where $\theta$ increases counterclockwise, see Figure 3.2). We know that that the constant $R = \sqrt{K/2\pi V_\infty}$ and at the cylinder's surface where $r = R$ Eqs. (3.14) and (3.15) become

$$V_r = V_\infty \cos\theta\left(1 - \frac{R^2}{R^2}\right) = 0 \tag{4.8}$$

$$V_\theta = -V_\infty \sin\theta\left(1 + \frac{R^2}{R^2}\right) - \frac{\Gamma_0}{2\pi R} = -2V_\infty \sin\theta - \frac{\Gamma_0}{2\pi R} \tag{4.9}$$

In order to find the pressure coefficient ($C_p$) at any given surface location with Eq. (4.5), we need the ratio $V_\theta/V_\infty$. To arrive at the lift force per unit span, we will then have to integrate $C_p$ over the entire cylindrical surface.

$$C_p = 1 - \left(\frac{V_\theta}{V_\infty}\right)^2 = 1 - \frac{1}{V_\infty^2}\left[4V_\infty^2 \sin^2\theta + \frac{2\Gamma_0 V_\infty \sin\theta}{\pi R} + \frac{\Gamma_0^2}{4\pi^2 R^2}\right] \tag{4.10}$$

To calculate the lift coefficient on the cylinder, we use a polar form of Eq. (2.31), changing $x$ to polar form with a cylinder of chord $c = 2R$, a constant value. The integral in terms of the variable $\theta$ now goes from 0 to $2\pi$. It can be shown that only even powers of the sine-function, namely, only the middle bracketed term in Eq. (4.10), will survive the integral of $C_p \sin\theta$ around the circle which is shown below.

$$c_\ell = -\frac{1}{2}\int_0^{2\pi} C_p \sin\theta \, d\theta = -\frac{1}{2}\int_0^{2\pi} \frac{2\Gamma_0 \sin^2\theta}{\pi V_\infty R} d\theta = \frac{\Gamma_0}{V_\infty R} \tag{4.11}$$

As depicted in Figure 4.3, there can be at most two stagnation points on the flow around a cylinder with circulation and these depend on the strength of $\Gamma_0$. When these points merge at $\theta_S = 270^\circ$, we achieve maximum circulation with a stagnation point attached on the cylinder and that condition is represented by the vortex strength $\Gamma_{0max} = 4\pi R V_\infty$. Inserting this value on the right-hand side of Eq. (4.11), the resulting lift coefficient at this maximum condition becomes

$$c_{\ell\, max} = 4\pi$$

In Chapter 11, we show that, beyond being a purely theoretical value, this maximum lift coefficient is not achievable even with ideal flow around slender airfoils. The cylinder lifting surface has no "angle of attack" dependence (i.e. $c = 2R$ in Eq. (2.20) under any cylinder orientations), whereas thin symmetric and other teardrop-shaped airfoils do.

Using Eq. (4.11), the lift per unit span ($\ell$) may now be written as

$$\ell = \frac{1}{2}\rho_\infty V_\infty^2 (2R) c_\ell = \rho_\infty V_\infty \Gamma_0 \tag{4.12}$$

Even though Eq. (4.12) was derived for a flow over a cylindrical object, this result is known not depend on any geometric details

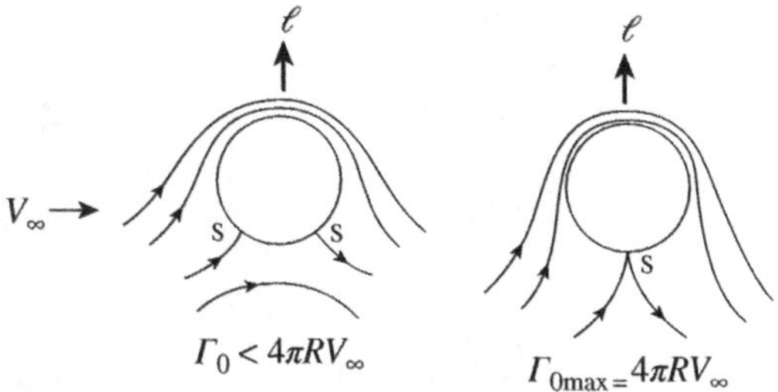


**Figure 4.3** Flow around a cylinder with different amounts of circulation.

of the lifting object (the use of a cylinder was for mathematical convenience). Rather, the lift per unit span depends only on the product of the free stream density with the flow velocity and the magnitude of the circulation at the object's surface. Equation (4.12) was arrived at independently by Kutta in Germany, Joukowski in Russia, and is a most important theoretical result in aerodynamics. It shows that for an airfoil to produce lift, circulation must be generated around it and, as we first mentioned in Chapter 1, this originates within a thin viscous region terminating in a stagnation point at the airfoil's trailing edge. Flow circulation, however, is not always clearly apparent around airfoils or even cylinders as streamlines do not seem to move *around* immersed objects since they are being simultaneously being convected downstream. The best way to show that a flow has circulation is to establish that Eq. (3.7) is not zero (as done in Example 3.3).

**Example 4.2** Show that uniform flow around a circular cylinder without circulation produces no net forces even under ideal conditions (i.e. that there is no lift or drag). Also, discuss the "yaw probe" application.

The superposition of a *uniform flow* approaching from the left with a *doublet* of strength $K$ in two dimensions has been shown to generate a circular cylinder. The tangential velocity $V_\theta$ at a cylinder's surface of radius $K/2\pi V_\infty \equiv R^2$ and the corresponding surface pressure coefficient at $r = R$ both turn out to be

$$V_r = V_\infty \left(1 - \frac{R^2}{r^2}\right) \cos\theta = 0 \quad \text{and} \quad V_\theta = -V_\infty \left(1 + \frac{R^2}{r^2}\right) \sin\theta = -2V_\infty \sin\theta$$

$$C_p = 1 - \left(\frac{V_\theta}{V_\infty}\right)^2 = 1 - 4\sin^2\theta$$

For this ideal flow, the resulting $C_p$ is symmetrical about both the vertical and horizonal planes as seen in Figure 4.2 so that any *net pressure* force is zero. This can also be shown by the fact that the second integral in Eq. (4.11) has an even-power sine term (see Problem 4.4).

The pattern of the pressure shown in Figure 4.2 is only realistic on the upstream side because on the back side the flow separates shortly after the lowest pressure at the peak. However, the distribution of pressure on the cylinder's front side can be used as a relative wind-direction indicator or "cylindrical yaw probe" by placing one hole at $\theta' = 0^\circ$ and 2 holes at $\theta' = \pm 30^\circ$ (where $4\sin^2\theta' = 1.0$), here $\theta' = 0^\circ$ when $x = 0$. Whenever the pressure of the two outer holes is equal, the pressure in the center hole is the stagnation pressure (this arrangement also acts as a stagnation pressure orifice where pressure measurements could be sensed in a Pitot-static manner). See Figure E4.1.

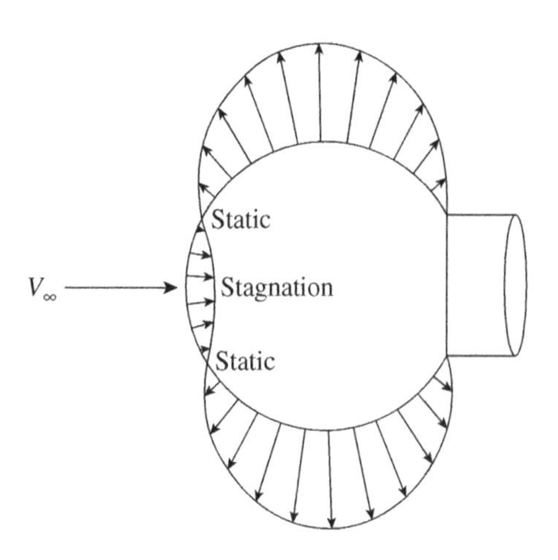


**Figure E4.1** Cylindrical yaw probe design with an instrumentation outlet port on the right end.

## 4.6 Pressure–Energy Equation

A common application of Bernoulli's equation is in the analysis of flows in internal passages of low-speed wind or water tunnels. This equation's utility may be expanded by writing it in combination with the First Law of thermodynamics (Zucker and Biblarz 2020), since in Eq. (4.3) each term has units of energy per unit mass (e.g. J/kg). When mechanical or *shaft-work* per unit mass ($w_{s1-2}$) as well as *frictional loss* terms are added to the sum of the flow's kinetic and potential energies per unit volume, Eq. (4.4) divided by the constant $\rho$, becomes

$$\frac{p_1}{\rho} + \frac{V_1^2}{2} + gz_1 = \frac{p_2}{\rho} + \frac{V_2^2}{2} + gz_2 + w_{s1-2} + \text{losses} \tag{4.13}$$

The shaft work term may represent a pump or other mechanical component (which is negative if work is done *on* the fluid), and the loss term is usually related to wall friction (which is always positive). Even though heat transfer does not appear in Eq. (4.13), it can be shown to apply to non-adiabatic cases.

## 4.7 Enrichment Topics

Note: Starting with this chapter, we introduce one or more topics related to but not traditional to basic aerodynamics. It is hoped that these topics will help stimulate the reader and provide a better understanding of the subject.

### 4.7.1 Aero-Lift-Force Electromagnetic Analogy

In order to compare aerodynamic and electromagnetic forces, we need to examine the total lift by integrating Eq. (4.12) over the wing's span. In vector form, total lift on a wing becomes

$$\vec{L} = \int \left(\rho_\infty \vec{V}_\infty \times \Gamma d\vec{y}\right) \quad \text{(N)} \tag{4.14}$$

We need to only identify the Cartesian unit-vector direction of $V_\infty\left(\vec{i}\right)$, $y\left(-\vec{j}\right)$ and $L\left(\vec{k}\right)$, since $\rho_\infty$ and $\Gamma$ are scalars. You might find Eq. (4.14) unusual because the resultant lift force is perpendicular to the two vectors that generate it – the free stream velocity vector $\vec{V}_\infty$ and the vorticity directed along the span – and this is unlike both gravity and electrostatic forces that act along a direction connecting the masses under scrutiny. The aerodynamic force shares this unique characteristic with the electromagnetic or "Lorentz force," $\left(q_e\vec{V}_e \times \vec{B}\right)$ (N), where $q_e$ is electric charge, $\vec{V}_e$ its velocity, and $\vec{B}$ the magnetic induction. Note that $q_e\vec{V}$ represents *moving charges* i.e., an electric current (which can be shown to satisfy mass continuity, Eq. (2.8)), and that $\vec{B}$ comes from an external or imposed magnetic field.

The analogy between electromagnetism and fluid mechanics was recognized shortly after Maxwell's equations were formulated and this equivalence is closely examined in Example 4.3. Devices based on of Lorentz force can be found in the so-called "*J*×*B* accelerators" also known as magnetoplasma dynamic (MPD) accelerators (as shown in Figure 4.4) and in pulsed-plasma thrusters (PPT) (Sutton and Biblarz 2017).

**Example 4.3** Formulate the lift force on a wing of finite span and compare it with the total Lorentz force produced by a magnetic field oriented as shown in Figure 4.4 which acts on a current flowing in the *y*-direction. Note that from electromagnetism, $\vec{B} = \mu\vec{H}$ where $\vec{H}$ is the magnetic field intensity and that for a steady-state, the current density $\vec{J} = \nabla \times \vec{H}$.

The following equations compare these two total forces in the SI-system of units. In Eq. (E4.3.1), we have used Eq. (3.8) to replace the circulation and this leads to the volume integral shown. Noting further that the Lorentz force is a force per unit volume we have, ($\mathbb{V}$ = control volume)

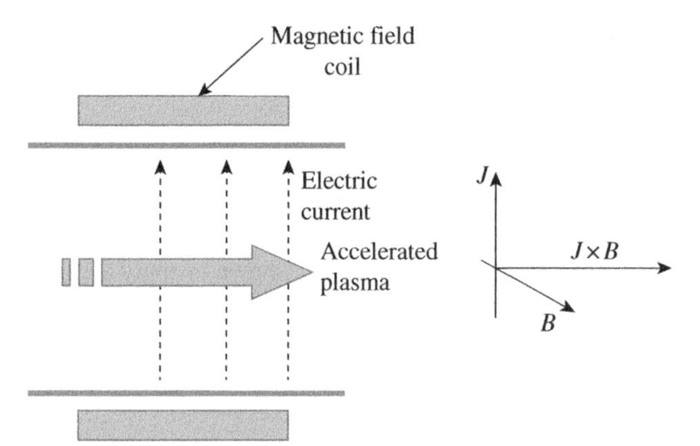


**Figure 4.4** MPD plasma accelerator. Plasmas are ionized gases that conduct electricity.

$$\text{Aerodynamic force: } \vec{L} = \int \rho\vec{V} \times \left(\Gamma d\vec{y}\right) = -\int \rho\vec{V} \times \left(\nabla \times \vec{V}\right) d\mathbb{V} \tag{E4.3.1}$$

$$\text{Electromagnetic force: } \vec{F} = \int \vec{J} \times \vec{B} d\mathbb{V} = \int \nabla \times \vec{H} \times \mu\vec{H}\, d\mathbb{V} = -\int \mu\vec{H} \times \left(\nabla \times \vec{H}\right) d\mathbb{V} \tag{E4.3.2}$$

Both forces above are in (N), $\mu$ is in (N/A$^2$), and $H$ in (A/m). In free space, $\mu = \mu_0 = 4\,\pi \times 10^{-7}$ N/A$^2$, where the symbols mean (A) ≡ Ampere and (N) ≡ Newton.

In equation (E4.3.2), $\mu$ is the permeability of the ionized moving medium and we have used a *Maxwell equation* to relate the current density to the magnetic field intensity $\vec{H}$ for a steady state. When the ionized medium is non-magnetizable, $\mu$ becomes the "permeability of free space" which is a constant and when the flow is incompressible the fluid density is taken as constant. This comparison shows that the aerodynamic and Lorentz forces have equivalent equation forms so that the fluid velocity field $\vec{V}$ is analogous to the magnetic field intensity $\vec{H}$. Moreover, circulation is shown below to be analogous to the electric current. These correspondences may be visualized with a magnetic field produced by a solenoid wound in an appropriate direction (see Figure E4.3). The negative signs in the equations above arise because by convention the positive sense of a line integrals is counterclockwise and for a fluid flow approaching from the left the circulation has to be clockwise to represent lift (see Section 3.4).

Next, using the analog between $\vec{V}$ and $\vec{H}$, we can apply another steady-state Maxwell equation to find the electrical analog of the circulation ($\Gamma$). For this, we assume that the contour integrals below exist in a plane perpendicular to the directions of $V$ and $H$ ($\mathbb{A} =$ control surface area).

$$-\Gamma = \oint_C \vec{V} \cdot d\vec{r} \quad \text{and} \quad \oint_C \vec{H} \cdot d\vec{r} = \int_S \vec{J} \cdot d\vec{\mathbb{A}} = I \tag{E4.3.3}$$

$$\Gamma \ \left(\mathrm{m^2/s}\right) \quad \leftrightarrow \quad I \ (\mathrm{A})$$

It can also be shown from Eqs. (E4.3.1) and (E4.3.2) that we can apply another of Maxwell's equations and arrive at the steady form of the continuity, Eq. (2.7). This reinforces the results already found earlier.

$$\nabla \cdot \vec{B} = \nabla \cdot \mu\vec{H} = 0 \quad \text{and} \quad \nabla \cdot \rho\vec{V} = 0 \tag{E4.3.4}$$

$$\vec{H} \ (\mathrm{A/m}) \quad \leftrightarrow \quad \vec{V} \ (\mathrm{m/s})]$$

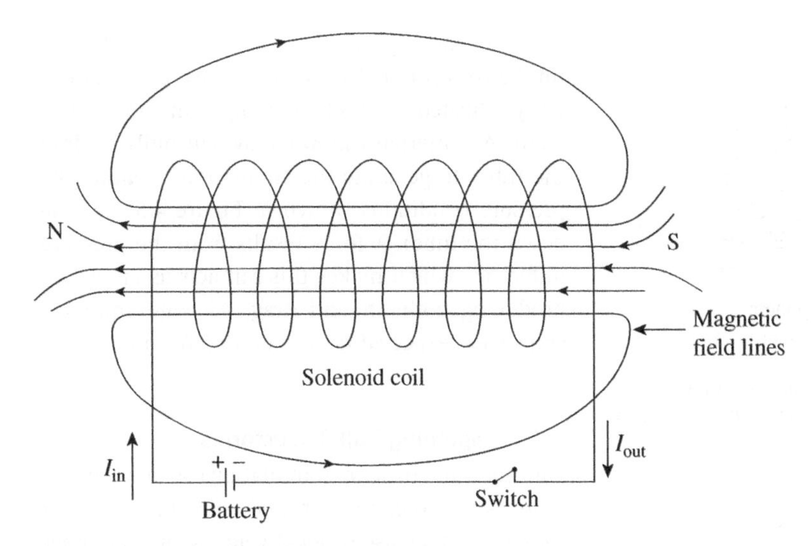


**Figure E4.3** Solenoid windings producing a constant magnetic field inside the coil. Here the analogy between the current in the solenoid and a fluid mass rotating as a vortex filament generating vortex flow becomes apparent.

These electromagnetism fluid-dynamic analogs are also directly evident in the formulation of the induced velocity from a vortex filament as given by the Biot–Savart law.

### 4.7.2 Turbo-Sail Applications to Watercraft

The *Magnus effect* is a name given to the force on a spinning body immersed in a moving airstream; this force acts perpendicular to both the direction of the airstream and of the axis of rotation. Magnus tubular rotators have been used to propel ships and are called "rotor sails" because when the rotor is mounted with its axis vertical, this effect creates forward thrust in winds blowing horizontally. Thus, as with any sailing ship, a rotor ship moves forwards only when there is sufficient wind. Two different watercraft types with turbo-sails have been constructed and operated, namely, Flettner's Rotor Ship and Cousteau's Alcyone.

| *Web reference sites*: | https://en.wikipedia.org/wiki/Rotor_ship | Rotor or "Flettner rotor ship" |
|---|---|---|
| | https://en.wikipedia.org/wiki/Turbosail | Cousteau's Alcyone |

Flettner rotors consist of large, conventional, and vertical cylindrical structures installed on the deck of a ship, often with horizontal disc-shaped blades at the top and also sometimes down their length. They are rotated around a mast when there is sufficient wind present creating thrust from the Magnus effect. Rotor sails have reported 5–25% fuel savings but presently are not commonly used.

Cousteau's propulsion system has a sail-like tubular unit with a suction powered boundary-layer control system that can enhance the force generated across a wide range of angles of attack. The turbo-sail consists of an airfoil-like vertical and somewhat oval-shaped tube with a mobile flap (see Figures 4.5 and 4.6). This arrangement allows such a system to power boats in any direction simply by moving a single flap at the back of the sail, unlike conventional sails which have to be continually adjusted to react to changes in the relative wind. An internal *aspiration system* pulls air into the tubes to generate the circulation, creating the pressure conditions shown in Figure 4.5 (looking down the tube). A force results from the pressure difference which makes this sail acts as an airfoil, producing both lift and drag. Alcyone equipped ships were expected to save up to 50% in fuel.

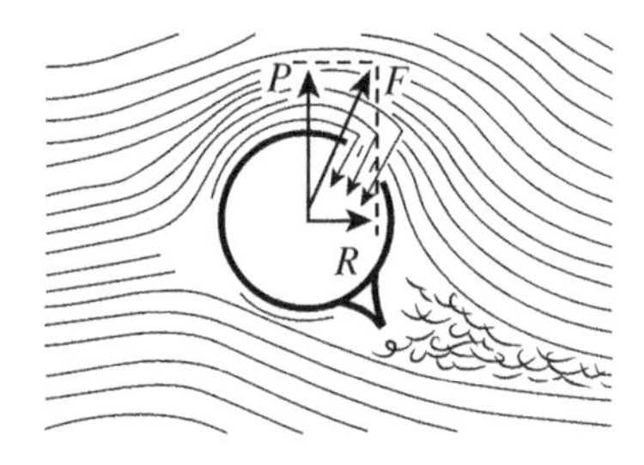


$P$ = Propulsive force component
$R$ = Resistive force component

**Figure 4.5** Alcyone generates circulation without rotation by aspirating the boundary layer.

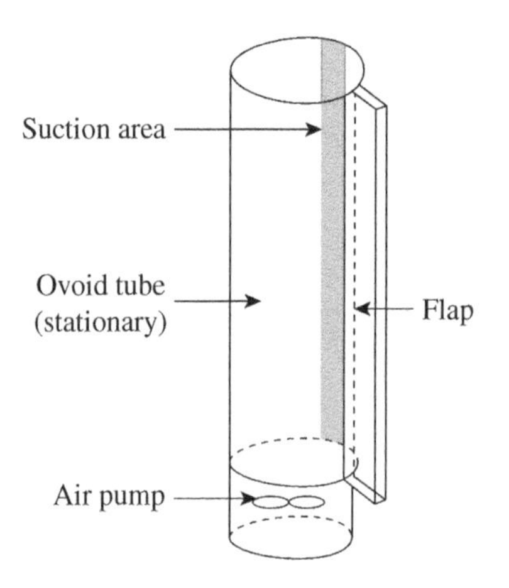


**Figure 4.6** Oval-shaped vertical sail on Alcyone. The cylinder's flap can be moved to align the propulsive force in any desired direction.

### 4.7.3 Spinning Ball Trajectories

The addition of spin (circulation) also produces changes of pressure around moving *spherical objects*. An increase of local velocity on the upper surface due to spin causes an equivalent increase of upper-surface suction, while the accompanying decrease in velocity on the lower surface causes a comparable decrease of lower surface suction. As a result, any moving spherical object with circulation will experience either a lift or a side force depending on the direction of its vorticity vector. Such result also comes from the Magnus effect and has the same relationship between aerodynamic circulation and lift, one that is important to golfers, baseball, and tennis players, as well as to pilots. The curvature of the flight path of a golf ball or baseball arises from unbalanced forces that are created by the rotation of the moving ball. Unstalled airfoils produce copious amounts circulation generating lift more efficiently and with less drag than rotating balls. This is because in airfoils the rear stagnation point is always pegged at their sharp tailing edge so there results less boundary layer separation.

## 4.8 Summary

Integrating Euler's equation along a streamline in incompressible, irrotational flows yields Bernoulli's equation. The symbol $p_t$ is used to represent the stagnation pressures that results when flows are isentropically decelerated to zero speed.

$$\boxed{\left(\frac{V_1^2}{2}\right) + \frac{p_1}{\rho} = \left(\frac{V_2^2}{2}\right) + \frac{p_2}{\rho} = \frac{p_t}{\rho}} \tag{4.3}$$

The Kutta–Joukowski theorem associates the *lift per unit span* to a bound vortex of strength $\Gamma_0$ in incompressible flows. Vorticity generated inside the boundary layers is continuously adjusted by the sharp trailing edge of an airfoil that makes it a stagnation point at angles of attack up to boundary layer separation.

$$\boxed{\ell = \rho_\infty V_\infty \Gamma_0} \tag{4.12}$$

In Example 4.3, it is shown that a velocity vector is analogous to a similarly directed magnetic field intensity and that fluid circulation is analogous to the flow of an electric current. These resemblances are verified not only from equivalent mathematical formulations but also with several well-known thrust producing mechanisms. Recognizing the relationship between electromagnetism and aerodynamics affirms use of the *Biot–Savart law*'s in calculating flow velocities associated with three-dimensional vortex filaments. The so-called Magnus effect has been used in marine propulsion and can have noticeable effects on the trajectories of spinning balls.

## Problems

**4.1** The "weight" of the static atmospheric is reflected in the pressure entries of the Standard Atmosphere table found in Appendix A. Using the form of equation (4.4) for Bernoulli's equation, calculate the magnitude of $\rho g \Delta z$ between sea level and 1200 m and compare your result to the change in pressure $\Delta p$ tabulated at each elevation. Because Bernoulli's equation describes a constant density fluid and air's density changes with elevation, use the density value tabulated in Appendix A at the average height (600 m) and for $g$ use 9.81 m/sec$^2$.

**4.2** Starting with the relation for the lift per unit span on a cylinder of radius $R$ given below, arrive at the form of the lift coefficient using appropriate definitions.

$$\ell = -\int_0^{2\pi} p \sin\theta R d\theta$$

**4.3** When ideal incompressible flow on a circular cylinder *approaches from above*, the form of the pressure coefficient at $r = R$ without circulation becomes as shown (here the polar coordinate axes remain located along the $x$-axis as before).

$$C_p = 1 - 4\cos^2\theta$$

Showing sketches, identify on the cylinder the stagnation points, maximum velocity locations as well as angular locations where the pressure is ambient.

**4.4** Given the three different pressure coefficients shown below, calculate the corresponding sectional lift coefficients effective on the cylinder. Show sketches. Note: $k$ = constant and use the first integral form in equation (4.11).

i) $C_p = -k$
ii) $C_p = -k\sin\theta$
iii) $C_p = -k\sin^2\theta$

**4.5** Measurements of the flow around an airfoil indicate that upstream the velocity, pressure, and density are 150 km/h, 0.91 bar, and 1.12 kg/m$^3$, respectively. If the lowest pressure reading at a particular location over the airfoil is 0.88 bar, what would be the velocity at that location? Is this value a maximum or a minimum? Assume ideal, constant density conditions.

**4.6** Pitot-static probe measurements over a flat plate in a wind tunnel are given below. Assume that $V_\infty = 30$ m/s, $\mu = 1.8 \times 10^{-5}$ kg/m-s, and $\rho = 1.2$ kg/m$^3$. The probe is moved by traversing the boundary layer in the *y*-direction with equal increments at a fixed *x*-location. Calculate the corresponding local velocities and plot them versus *y*.

| *y* (location #) | $p_t$ ($10^5$ N/m$^2$) | *p* ($10^5$ N/m$^2$) |
|---|---|---|
| 62 | 1.0187 | 1.0132 |
| 59 | 1.1085 | " |
| 56 | 1.0166 | " |
| 53 | 1.0143 | " |
| 50 | 1.0132 | " |

**4.7** A submarine moves through water at a depth of 60 feet at a speed of 30 ft/sec.
a) What is the hydrostatic pressure at that depth? Density of water $\rho = 62.4$ lbm/ft$^3$ at STP.
b) The speed of the submarine relative to the water is 45 ft/sec. Determine the effective total pressure on the submarine.

**4.8** The following values pertain to the same incompressible flow at two stations along different streamlines:

$$V_1 = 150\ \text{m/s},\ p_1 = 0.8 \times 10^5\ \text{N/m}^2, \quad \text{and} \quad V_2 = 60\ \text{m/s},\ \ p_2 = 1.013 \times 10^5\ \text{N/m}^2$$

Giving numerical values, indicate if this flow is rotational.

## Check Test

**4.1** In some flows, the so-called "Bernoulli constant" may vary between streamlines. What causes this?

**4.2** Does the relation $C_p = 1 - (V/V_\infty)^2$ ignore $\rho gz$?

**4.3** A light aircraft indicates (EAS) an airspeed of 266 km/h at a pressure altitude of 2400 m. If the outside air temperature is – 10 °C, what is the true airspeed (TAS)?

**4.4** Does a free counterclockwise vortex filament exposed to a relative wind that is approaching from its top experience lift?

**4.5** Show that Eq. (4.4) becomes Eq. (E4.1) in Example 4.1. Note that the height difference of the liquid $h_w$ is related to the pressure difference reading appropriately ratioed by different densities.

**4.6** Work Problem 4.5.

# 5

# Thin Airfoils in Two-Dimensional Incompressible Flow

## 5.1 Introduction

Thin airfoil theory can be used to model the behavior of many common wings. As long as the boundary layers remain attached and the Kutta condition is in effect, pressure force distributions around an airfoil accurately represent lift. Under these conditions, sectional drag originates and remains inside the boundary layers and the wake. As we saw in Chapter 3, vortex flow added to uniform flow are the ingredients that generate lift. In this chapter, we extend the potential vortex to a three-dimensional *vortex filament*; this is formally done using a formulation from electromagnetism. An array of closely packed line vortices can be constructed on an airfoil so as to form a *vortex sheet* that models distributed circulation around the airfoil. This vortex sheet which now embodies the physical airfoil is positioned along the airfoil's chord line or at a midpoint in cambered airfoils. At this location it acts as a "slip plane" because while flow streamlines coincide with the top and bottom profile of this surface they flow at much different velocities during lift production. Under thin airfoil theory assumptions, we may replace an arbitrary airfoil section with three additive parts: (a) a flat plate at angle of attack ($\alpha$), plus (b) a cambered plate at zero $\alpha$, together with (c) a symmetrical airfoil of the same thickness distribution about a mean camber line at zero $\alpha$. The flat plate part will contribute lift in proportion to its angle of attack and the camber curvature will contribute a fixed amount of lift as a function of its shape at zero angle of attack. The symmetric thickness or portion (c) contribution to the lift is usually neglected (but does influence the drag). See Figure 5.1 for details of this model.

In order for our sectional-characteristic formulations be extended to three-dimensional or finite wings, existing spanwise air flow must be small – this is generally the case in all but the shortest span wings. Effects of aft-end wing devices such as flaps will be incorporated by treating the flap as a variable-camber airfoil under the same theoretical approach.

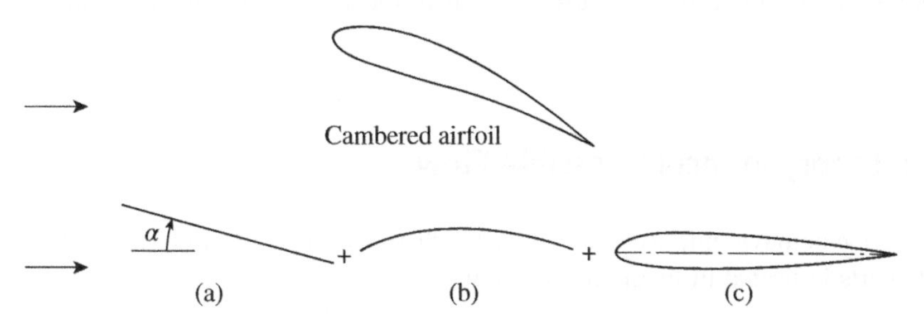


**Figure 5.1** Thin cambered airfoil separated into three contributions.

*Elements of Aerodynamics: A Concise Introduction to Physical Concepts*, First Edition. Oscar Biblarz.

Companion website: www.wiley.com/go/elementsofaerodynamics

## 5.2 Objectives

After completing this chapter, you should be able to:

1) Describe the physical character of a vortex filament and explain how the superposition of a two-dimensional vortex filament with a cross-flowing free stream produces lift.
2) Separate the thin airfoil model into its three components and identify those parts that contribute to the lift.
3) Explain how the integral of the line-distributed vorticity $\gamma(x)$ over a chord or camber line represents total circulation and thus the lift per unit span on thin airfoils.
4) Identify all conditions under which the equation $c_\ell = 2\pi\alpha$ applies.
5) Explain why the center of pressure is always located at the quarter-chord in subsonic symmetric airfoils.
6) Identify the conditions under which the equation $c_\ell = 2\pi(\alpha - \alpha_{\ell 0})$ applies and give meaning to the zero-lift angle of attack ($\alpha_{\ell 0}$).
7) Describe what aft-end flaps best accomplish and why they may be treated as airfoils with adjustable camber under our formulations.

## 5.3 The Vortex Filament

In electromagnetics, the Biot–Savart law gives the intensity of an electromagnetic field around a conductor in terms of the magnitude of the current flowing through it. In aeronautics, it is used to establish the relationship between a circular velocity ($V_\theta$) induced by a vortex filament and the strength ($\Gamma_0$) or circulation of the vortex filament at any radial location.

For a straight, unbounded vortex filament, the induced circumferential velocity is given by,

$$V_\theta = \frac{\Gamma_0}{2\pi r} \tag{5.1}$$

where the polar coordinates $(r, \theta)$ are fixed with the cross-section of the filament. In other words, the potential vortex introduced in Chapter 3 represents a "slice" of the vortex filament itself.

A vortex filament has vorticity only at its core so that any circulation around a closed path that *excludes* the origin is zero, and this is how we may construct an *irrotational region* around such filament or collection of filaments in order to apply Bernoulli's equation to flows on lifting airfoils. We should expect that when boundary layers are separated and turbulent, this model breaks down and we are unable to calculate the pressures that surround the airfoil with Bernoulli's equation.

## 5.4 Thin Airfoil Theory in Incompressible Flow

Our goal here is to describe aerodynamic sectional characteristics of thin airfoils in terms of two of the three aggregate parts indicated in Figure 5.1, namely,

a) a flat plate at angle of attack $\alpha$.
b) a cambered plate at zero angle of attack.

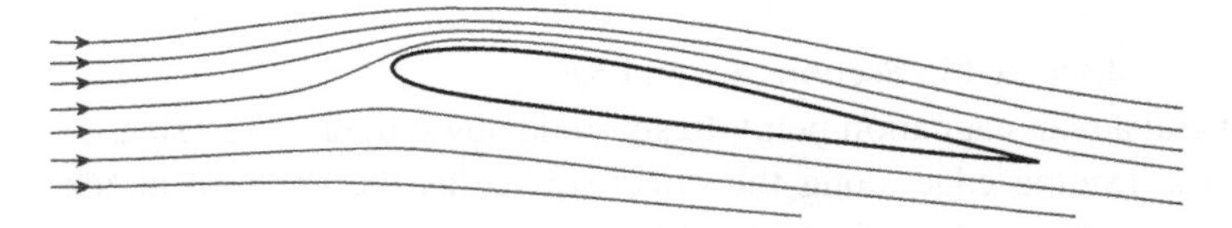

**Figure 5.2** Thin airfoil showing steamlines beyond the boundary layers.

A symmetric airfoil's thickness at zero angle will not measurably contribute to the lift and has only a small effect on angle of attack geometry, but an appropriate thickness distribution is always necessary for boundary layer management (see Chapter 7). In Figure 5.2, we show a thin airfoil under ideal flow that has been configured by the Kutta condition and we will study vorticity effects as if they were only present at its mean line.

The separate parts depicted in Figure 5.1 will be restricted to a maximum-thickness to chord ratios ($t_m/c$) of 15%, to angles $\alpha$ below 20° so that flows remain attached, and to Mach numbers below 0.3 for incompressibility. In our analysis, we use three-dimensional Cartesian coordinate symbols ($x$, $y$, $z$) in our formulations, where $x$ is along the chord, $y$ along the span, and $z$ in the vertical or cross-flow direction.

In Chapter 4, we arrived at the expression of the lift per unit span given as

$$\ell = \rho_\infty V_\infty \Gamma \tag{4.12}$$

which comprises both the relative velocity and the circulation needed to generate lift. But to describe the entire airfoil section, we have to expand the vortex filament concept to a *vortex sheet* located along a flat plate at an angle of attack $\alpha$, and for cambered airfoils also along the camber line at zero $\alpha$. This is accomplished by defining $\gamma(x)$, a variable circulation per-unit-chord-length that may be added (or integrated when continuously distributed) along the $x$-direction to generate the total circulation ($\Gamma$). Figures 5.3 and 5.4 illustrate a series of discrete vortex filaments spread

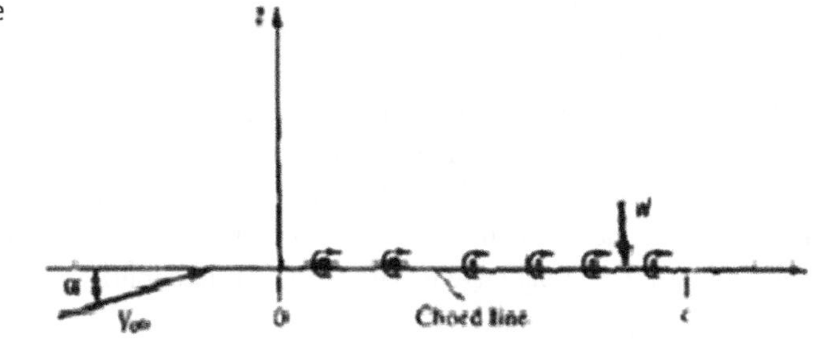


**Figure 5.3** Symmetric airfoil at angle of attack $\alpha$, portion (a) in Figure 5.1.

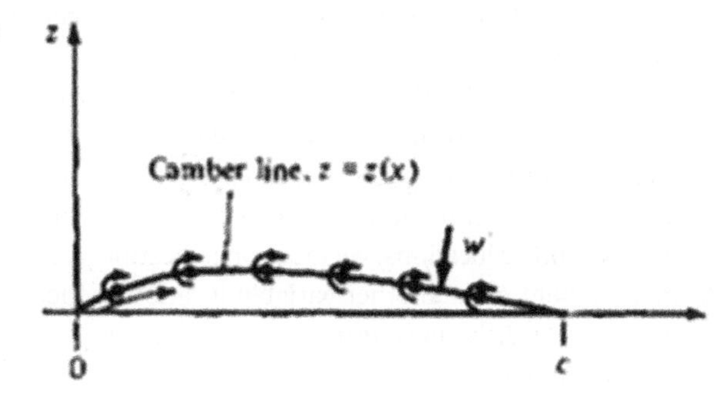


**Figure 5.4** Cambered airfoil portion at zero $\alpha$, portion (b) of Figure 5.1.

out along the mean-lines in both types of airfoil. These figures also show an induced velocity ($w$) normal to each mean line and it is discussed in Sections 5.5 and 5.6.

Airfoils have both a forward and a rear stagnation point. In subsonic flows, the forward stagnation point moves away from the leading edge along the lower surface as the angle of attack increases, and the flow going toward the upper surface after this stagnation point needs to be kept attached by the smoothness of the airfoil's leading edge (and also with the help of leading-edge deployable surfaces at high angles of attack).

## 5.5 Symmetric Contribution at Angle of Attack

We will now examine the contribution from a thin flat plate at angle of attack, portion (a) in Figure 5.1. The lift generated by each vortex filament present must be "summed" or, since these vortices are continuously distributed, integrated as in Eq. (5.2) where

$$\Gamma = \int_0^c \gamma(x)dx$$

We can then form the sectional lift coefficient as in Eq. (5.3). With such line-distributed vorticity, $\gamma(x)$ may be further related to the local pressure difference, ($p_l$) acting under the plate and ($p_u$) above it as indicated in the second equal sign of Eq. (5.3). We have used Eq. (2.31) for the equivalence shown in terms of the $C_p$s.

$$\ell = \rho_\infty V_\infty \int_0^c \gamma(x)dx \tag{5.2}$$

$$c_\ell = \frac{\ell}{1/2\rho_\infty V_\infty^2 c} = \frac{2\int_0^c \gamma(x)dx}{V_\infty c} = \int_0^1 (C_{pl} - C_{pu})d\left(\frac{x}{c}\right) \tag{5.3}$$

$$\gamma(x) \equiv \frac{p_l - p_u}{\rho_\infty V_\infty} \tag{5.4}$$

Equation (5.4) comes from the proportionality of the distribution of vorticity $\gamma(x)$ to the local pressure difference across the plate ($p_l - p_u$) that make up $C_p$. All of the above results apply to both cambered and symmetric thin airfoils. Since flow lines cannot cross a surface, the streamlines above and below the mean line must be parallel to it. Moreover, upper and lower flow paths must merge smoothly at the trailing edge to satisfy the Kutta condition, i.e.

$$\gamma(c) = 0$$

Recall that with real flows this requires an airfoil configuration where the effects of viscosity are sharply focused at the its trailing edge.

In order to solve for the vorticity distribution surrounding a flat plate, we use the fact that the resulting *induced vertical velocity* ($w$) (shown in Figure 5.3) must cancel the vertical portion of the free stream velocity ($V_\infty \sin\alpha$) at each $x$-location along the plate and at each angle of attack $\alpha$ because there can be no flow across the airfoil. Since all points along the vortex sheet induce velocities at other locations, we focus on a location $x_0$ along the plate and examine the contribution from a segment $dx$ at another variable location $x$. Including $\gamma(x)$ and having been modified with $r = (x_0 - x)$, Eq. 5.1 as next written along the $x$-axis includes all vortex contributions to the induced vertical velocity $w(x_0)$

$$dw = -\frac{\gamma(x)dx}{2\pi(x_0 - x)} \tag{5.5}$$

$$w(x_0) = -\frac{1}{2\pi}\int_0^c \frac{\gamma(x)dx}{x_0 - x} \tag{5.6}$$

Since this induced velocity must be equal and opposite to the total vertical components of the free stream, we can write

$$\frac{1}{2\pi}\int_0^c \frac{\gamma(x)dx}{x_0 - x} = V_\infty \sin\alpha \tag{5.7}$$

*This is the fundamental equation of thin airfoil theory.* Equation (5.7) needs to be solved for the flat plate distribution of $\gamma(x)$, even though it appears inside the integral. There is a well-known classical solution to such an *integral equation* in of Cartesian coordinates form given as Eq. (5.8) (the polar coordinate version is found in aerodynamics books, e.g. Anderson 2017; Bertin and Cummings 2013; Glauert 1947; Houghton and Carpenter 1993; Katz and Plotkin 2000; Kuethe and Chow 1998; McCormick 1979; Prandtl and Tietjens 1934a, b; Schlichting and Truckenbrodt 1979; Shevell 1983). After integrating over $x_0$ we get the desired end result for the total sectional lift and its coefficient in Eqs. (5.9) and (5.10a),

$$\gamma_s(x_0) = 2V_\infty \sin\alpha\sqrt{\frac{c - x_0}{x_0}} \tag{5.8}$$

$$\ell = \rho_\infty V_\infty \sin\alpha\int_0^c \gamma_s(x_0)dx_0 = \pi\rho_\infty V_\infty^2 c\sin\alpha \tag{5.9}$$

$$\boxed{c_\ell = 2\pi\sin\alpha \approx 2\pi\alpha} \tag{5.10a}$$

$$\boxed{c_{\ell\alpha} \approx 2\pi} \tag{5.10b}$$

We have implicitly assumed that the angles of attack are small enough to apply the linear approximation $\sin\alpha = \alpha$ and will continue to use it up to Chapter 10. In Chapter 11, we revisit situations for angles of attack above 20°. Equations (5.10a,b) show that for a thin symmetric airfoil, the sectional lift coefficient is proportional to the angle of attack with a slope of around $2\pi$ and this result has been extensively verified experimentally.

### 5.5.1 Pitching Moment About the Leading Edge

The *pitching moment about the leading edge* ($m_0$) is formulated next. Figure 5.5 shows the associated vector diagram where a nose-down condition carries a negative sign.

$$m_0 = -\rho_\infty V_\infty\int_0^c \gamma(x)x dx = -\frac{\pi}{4}\rho_\infty V_\infty^2 \sin\alpha c^2 \approx -\frac{\pi}{4}\rho_\infty V_\infty^2 \alpha c^2 \tag{5.11}$$

$$c_{m0} \equiv \frac{m_0}{q_\infty c^2} = -\frac{\pi}{2}\alpha = -\frac{c_\ell}{4} \tag{5.12}$$

We see in Eq. (5.12) that whenever lift is present, symmetric airfoils have a pitching moment about the leading edge that is nose down and proportional to the sectional lift coefficient.

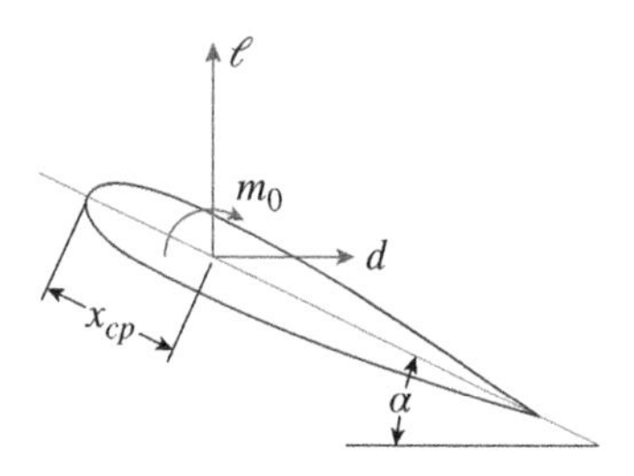


**Figure 5.5** Pitching moment ($m_0$) about leading edge with net vectors shown at the center of pressure ($x_{cp}$).

### 5.5.2 Center of Pressure and Aerodynamic Center

The *center of pressure* ($x_{cp}$) has been defined in Chapter 2 as the chord location about which the pitching moment is zero. From Figure 5.5, we need to solve

$$m_0 + x_{cp}\ell = 0$$

from Eqs. (5.9) and (5.11). The ratio $x_{cp}/c$ ultimately becomes

$$\frac{x_{cp}}{c} = \frac{c_{m0}}{c_\ell} \quad \text{or} \quad x_{cp} = c/4 \tag{5.13}$$

This result is independent of the angle of attack for flat plates and thus for any thin symmetric airfoil. It turns out, therefore, that the center of pressure ($x_{cp}$) and the location of moment independence on angle of attack, known as the *aerodynamic center* ($x_{ac}$) coincide for symmetric airfoils, i.e.

$$x_{cp} = x_{ac}$$

as depicted in Figure 2.4a. This is not true for cambered airfoils because the resultant upper surface's pressure force is at a greater chord location than that from the lower surface as shown in Figure 2.4b.

**Example 5.1** For a symmetric airfoil at an angle of attack of 12°,

a) plot the pressure difference coefficient $\Delta C_p$ as given in Eq. (E5.1).
b) what are the corresponding values of $c_\ell$ and $c_{mac}$?

a) A thin symmetric airfoil is replaced by a flat plate for which the desired resulting pressure difference and its corresponding coefficient may be written as (from Eqs.(5.4) and (5.8) with the small angle approximation)

$$p_l - p_u = \rho_\infty V_\infty \gamma(x) = 2\rho_\infty V_\infty^2 \alpha \sqrt{\frac{c-x}{x}}$$

$$\Delta C_p \equiv \frac{p_u - p_l}{q_\infty} = -4\alpha\sqrt{\frac{1-x/c}{x/c}} \tag{E5.1}$$

Note in Figure E5.1 that for the solution of this problem, the curve becomes very large at the leading edge – fortunately, this is a "manageable singularity" because the integral over the chord of $\gamma(x)$ remains finite. Furthermore, the curve goes to zero at the trailing edge where the upper and lower pressures become equal. It is clearly evident from the behavior of $\Delta C_p$ that pressure differences are higher at the first half of the plate (a feature of all subsonic flows).

b) At 12° (0.21 radians) the lift coefficient from Eq. (5.10a) is $c_\ell = 2\pi\alpha = 1.31$. Since for a flat plate the aerodynamic center is the same as the center of pressure, we find that here the moment coefficient is zero.

$$c_{mac} = c_{mcp} = c_{m0} + \frac{x_{cp}}{c} c_l = -\frac{c_\ell}{4} + \frac{c_\ell}{4} = 0$$

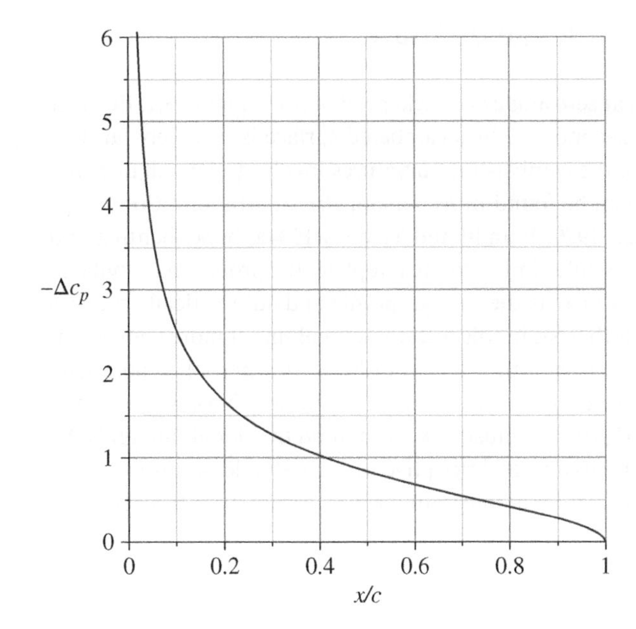


**Figure E5.1** Pressure coefficient difference as a function of $x/c$ for a symmetric airfoil at an angle of attack of 12°.

Figure 5.6 shows experimental results for a symmetric airfoil. When we calculate values of the *pressure coefficient difference* ($C_{pl} - C_{pu}$) and to plot them against $x/c$, we should get a graph very similar to Figure E5.1 except at the leading and trailing ends (see Problem 5.2).

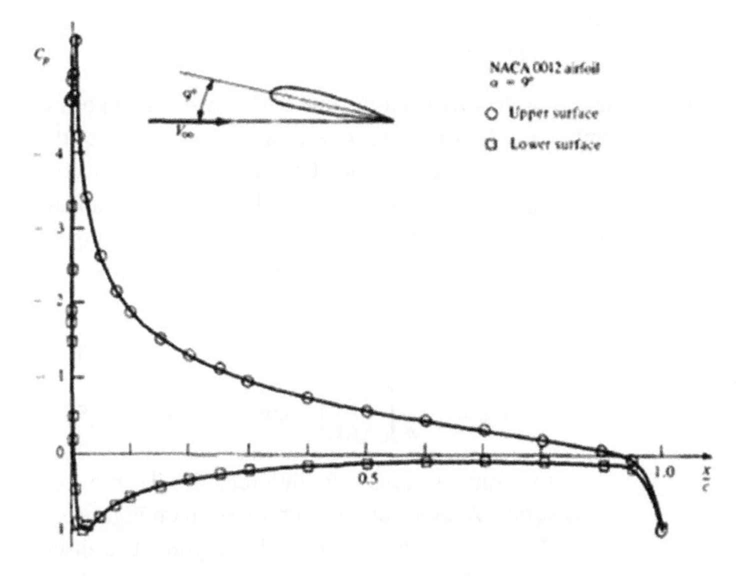


**Figure 5.6** Experimental verification of thin airfoil theory using a symmetric NACA 0012 airfoil at $\alpha$ = 9°. Circles are $C_p$ measurements for lower surface and squares $C_p$ measurements for upper surface vs $x/c$. Source: Data from Abbott and von Doenhoff (1949).

## 5.6 Camber Contribution at Zero Angle of Attack

We next examine the camber contribution at zero angle of attack, portion (b) in Figure 5.1. Even for thin airfoils the calculation for the velocities induced by a cambered surface is relatively burdensome due to the need to model specifics of the camber-line curvatures involved. We will dispense with much detail of the derivation (they can be found in Anderson, 2017; Bertin and Cummings 2013; Kuethe and Chow 1998; McCormick 1979; Prandtl and Tietjens 1934a, b; Schlichting and Truckenbrodt 1979; Shevell 1983) because, while similar to the flat plate, the proof is manipulated through the use of a *Fourier sine-series* that models the camber profile and automatically satisfies the Kutta condition, see Appendix C. Nevertheless, in this section, we will find it more convenient to express the slope of the camber line in terms of the polar coordinate $\theta$, where the coordinate transformation centers a circle at mid-chord ($c/2$) as given in Eq. (5.14). Although we will continue to denote the slope of the camber line as $(dz/dx)_w$, it must be kept in mind that for all integrals the variable of integration is $d\theta$ (not $dx$) so that the camber-descriptive variable should correspond to it. Note that at $\theta = 0, x/c = 0$ and at $\theta = \pi, x/c = 1.0$ so that integrating with $x/c$ from 0 to 1.0 becomes an integration from 0 to $\pi$.

$$x = c/2(1 - \cos\theta)$$
$$dx = c/2(\sin\theta d\theta) \tag{5.14}$$

The ensuing camber contributions are shown below. We label the coefficients as $\Delta c_\ell$ and $\Delta c_{m0}$ because they must be added to the flat plate contribution which is the sole airfoil portion at angle of attack, as depicted in Figure 5.1.

$$\Delta c_\ell = 2\left[\int_0^\pi \left(\frac{dz}{dx}\right)_w (\cos\theta - 1)d\theta\right] \tag{5.15}$$

$$\Delta c_{m0} = -\left[\int_0^\pi \left(\frac{dz}{dx}\right)_w \left(\cos\theta - \frac{\cos(2\theta)}{2} + \frac{1}{2}\right)d\theta\right] \tag{5.16}$$

What results from this analysis is a constant new geometric entity, one that may be cast in terms of an angular contribution and given the symbol $\alpha_{\ell 0}$. This angle represents a shift to the left for the entire airfoil's $c_\ell$ curve as shown in Figure 5.7. When $\alpha_{\ell 0}$ is added to the flat plate portion, Eq. (5.10a), it becomes Eq. (5.17). Equation (5.18) gives the equation for calculating $\alpha_{\ell 0}$. Also apparent in Figure 5.7 is that, in addition to the lift curves being displaced to the left thus contributing lift a $\alpha = 0$, cambered airfoils may also increase the maximum value of the lift coefficient.

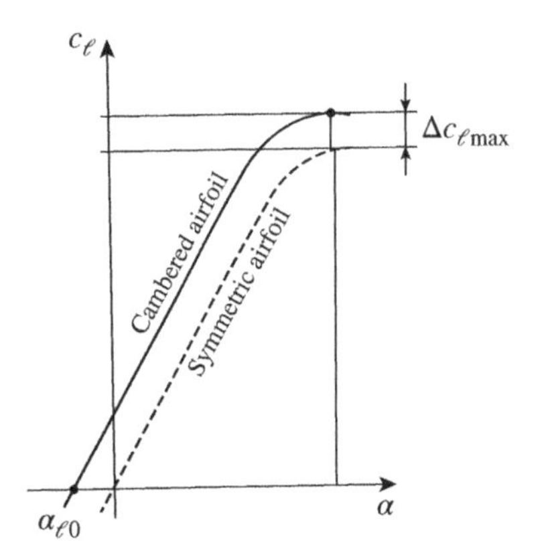


**Figure 5.7** Effect of camber on the angle of zero lift ($\alpha_{\ell 0}$) and on the maximum sectional lift coefficient ($c_{\ell max}$).

$$c_\ell = 2\pi(\alpha - \alpha_{\ell 0}) \tag{5.17}$$

$$\alpha_{\ell 0} = -\frac{1}{\pi}\int_0^\pi \left(\frac{dz}{dx}\right)_w (\cos\theta - 1)d\theta \tag{5.18}$$

Experimental results for different airfoil camber on several NACA airfoils are summarized on Figure 5.8.

While the pitching moment for a symmetric airfoil at its aerodynamic center is zero, on a cambered airfoil, it is negative or nose down as seen in Figure 2.4b because the net pressure force at the top of the airfoil is larger than that at the bottom. This figure is combined with airfoil data for the pitching moment and both are shown in Figure 5.9. The aerodynamic center location in cambered airfoils, however, remains

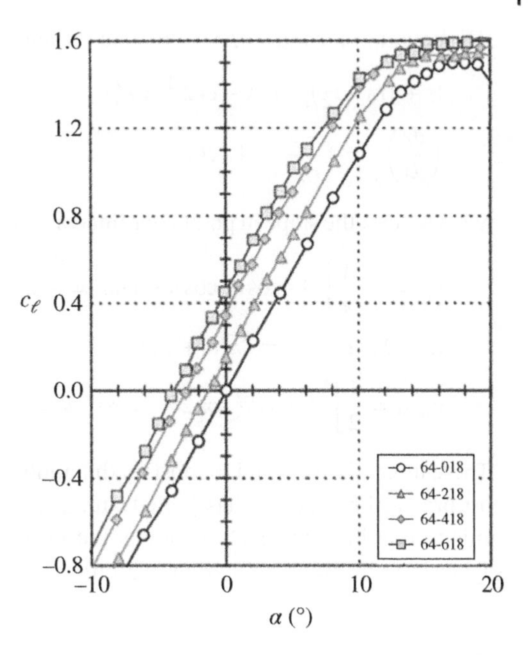


**Figure 5.8** Experimental results on 4 NACA 64-XXX airfoils. Source: Data from Abbott and von Doenhoff (1949).

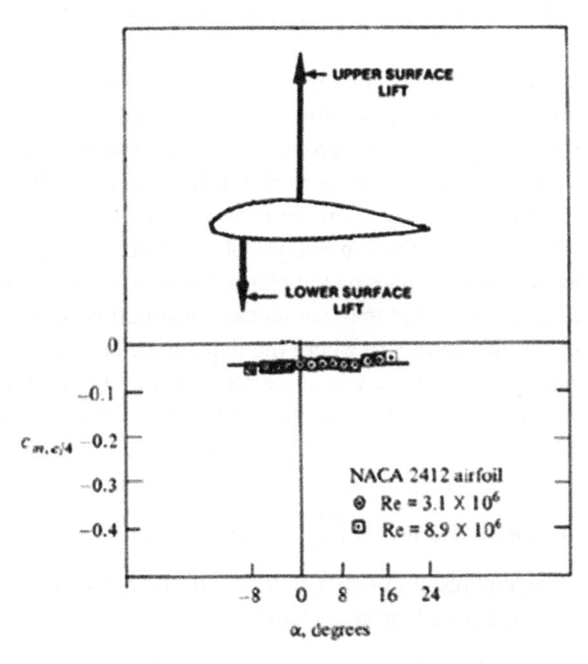


**Figure 5.9** Cambered airfoil data on $c_{mc/4}$. Source: Data from Abbott and von Doenhoff (1949).

at $c/4$ so there the pitching moment also remains independent of angle of attack. The moment coefficient about the aerodynamic center is given in Eq. (5.19).

$$\Delta c_{mac} = c_{mc/4} = \frac{1}{2}\int_0^{\pi}\left(\frac{dz}{dx}\right)_w(\cos 2\theta - \cos\theta)d\theta \tag{5.19}$$

**Example 5.2** The camber in a *circular-arc airfoil* is made up from the arc of a circle of radius $r_0$, chord $c$, and rise $\kappa c$, where $\kappa \ll 1$. Calculate $\alpha_{l0}$ and $c_{mc/4}$ at zero angle of attack for this cambered airfoil.

The equation for the camber line and its slope may be found from

$$[z + (r_0 - \kappa c)]^2 + [x - c/2]^2 = r_0^2$$

$$\left(\frac{dz}{dx}\right)_w = \frac{\frac{c}{2} - x}{r_0} = 4\,\kappa \cos\theta$$

We have cast the slope in polar coordinates as needed for the integrations in Eqs. (5.15) and (5.19).

$$\alpha_{\ell 0} = -\frac{1}{\pi}\int_0^{\pi} 4\,\kappa\cos\theta(\cos\theta - 1)d\theta = -2\kappa$$

$$c_\ell = 2\pi(\alpha - \alpha_{l0}) = 2\pi(\alpha + 2\kappa)$$

$$c_{mc/4} = \frac{1}{2}\int_0^{\pi} 4\kappa\cos\theta(\cos 2\theta - \cos\theta)d\theta = -\pi\kappa$$

These integrals are found in most math handbooks [check this!]. The circular arc airfoil is not the most practical profile but has the advantage of giving analytic results and we will return to this configuration in Chapter 11. Note that both calculations result in purely constant values and this is typical of any other camber configuration.

## 5.7 Flapped Symmetric Airfoil at Zero Angle of Attack

Careful scrutiny of the lift coefficient increase in Eq. (5.15) reveals that the portion of the mean camber line in the vicinity of the trailing edge has the biggest influence on $\alpha_{\ell 0}$. This is because the term $(1 - \cos\theta)$ is zero at the leading edge (where $\theta = 0$) and reaches its maximum value at the trailing edge (where $\theta = \pi$). It is on such effect the *flap* as a high-lift device is based (and also the *aileron* – a lateral aircraft control device).

We shall treat the flap as a type of thin airfoil configuration with *adjustable camber*. Here we need to only examine flaps on flat plates at zero angle of attack and because we assume that the flapping effect can be added to the other two contributions to the lift [what are they?]. The new geometry is shown in Figure 5.10 where the flap 1hinge is located at a distance $x_h$ from the leading edge and where the flap clockwise-angle rotation from the horizontal is $\eta$. We need to assume further that flap displacement does not appreciably change the chord length ($c$) of the airfoil which is shown as a dashed line in Figure 5.10 (even though it does at the higher values of $\eta$). The flap's hinge is at a fixed location $x_h$ with polar equivalent $\theta_h$ found from $x_h \approx 1/2c(1 - \cos\theta_h)$. The flap itself is flat so that for small $\eta$-displacements $\left(\frac{dz}{dx}\right)_w = \tan(\eta) \approx \eta$.

The end result of flap action on our aerodynamic parameters can be shown to be given by the following equations as $\Delta$-changes

$$\Delta c_{\ell f} = 2[(\pi - \theta_h) + \sin\theta_h]\eta \quad (5.20)$$

$$\Delta\alpha_{\ell f0} = -\frac{1}{\pi}[(\pi - \theta_h) + \sin\theta_h]\eta \quad (5.21)$$

$$\Delta c_{mfc/4} = 1/2[\sin\theta_h(\cos\theta_h - 1)]\eta \quad (5.22)$$

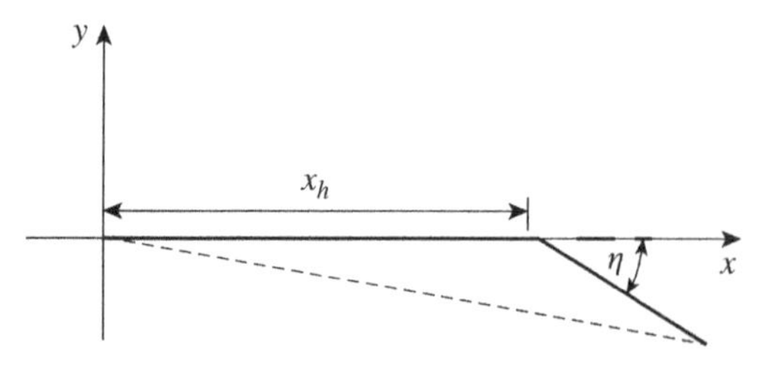


**Figure 5.10** Flap location on a symmetric airfoil

Figure 5.11 displays $\Delta c_{\ell f}$ results of using Eq. (5.20) for several values of $x_h$. It shows that flap contributions can reach lift coefficient values of the same order as an unflapped flat

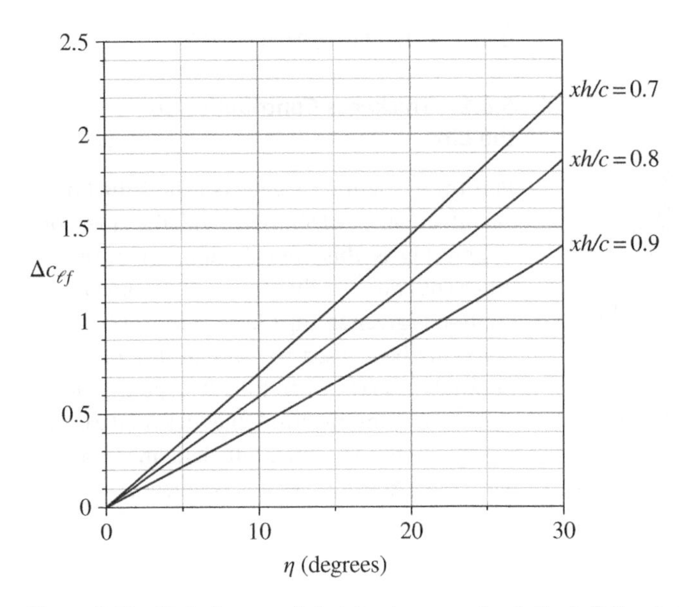


**Figure 5.11** Plain flap on a flat plate at zero angle of attack. Estimated $\Delta c_{\ell f}$ vs $\eta$ for typical values of $x_h/c$ using Eq. (5.20). Curves shown assume that $\theta_h$ does not vary with the flap angle $\eta$.

plate at 15° angle of attack. However, because of the many inherent approximations in their derivation, Eqs. (5.20), (5.21), and (5.22) do not represent flap behavior beyond $\eta > 30°$, where data indicate a certain maximum behavior.

Since activated flaps operate mostly within the airfoil's boundary layers instead of in ideal flows, these results do not faithfully represent observed trends with increasing flap angle. Another source of discrepancy with real flap behavior is the fact that our thin airfoil approximation is eventually exceeded at the higher flap angles. Figure 5.12 shows the effects of a 10° flap deflection as given by Eqs. (5.20) and (5.21).

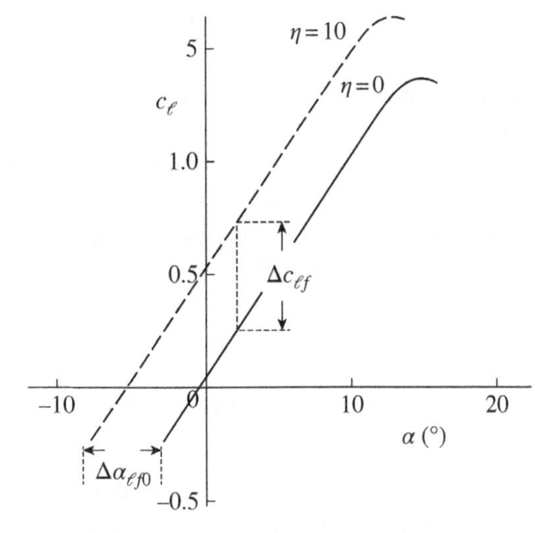


**Figure 5.12** Effect of flap deflection shown as $\Delta c_{\ell f}$ and $\Delta\alpha_{\ell f0}$

Airfoils are often equipped with leading edge flaps or *slats* along with regular flaps, but slats add insignificant amounts of lift or to $\Delta\alpha_{\ell 0}$; rather they are used for controlling boundary-layer separation at the airfoil's frontal region where major pressure differences are taking place. Slats can extend the range of linear operation by delaying the airfoil's stalling condition.

For regular flaps, Figure 5.13 shows a three-dimensional theoretical representation of $\Delta\alpha_{lf0}$.

The increase in $\Delta\alpha_{\ell f0}$ (in radians) due to flaps, Eq. (5.21), may alternatively be written in Cartesian coordinates without approximating to small flap angles as:

$$\Delta\alpha_{\ell f0} = \left[\left(1 - \frac{\cos^{-1}(1 - 2x_h/c)}{\pi}\right) + \frac{2\sqrt{x_h/c - (x_h/c)^2}}{\pi}\right] \tan(\eta) \tag{5.23}$$

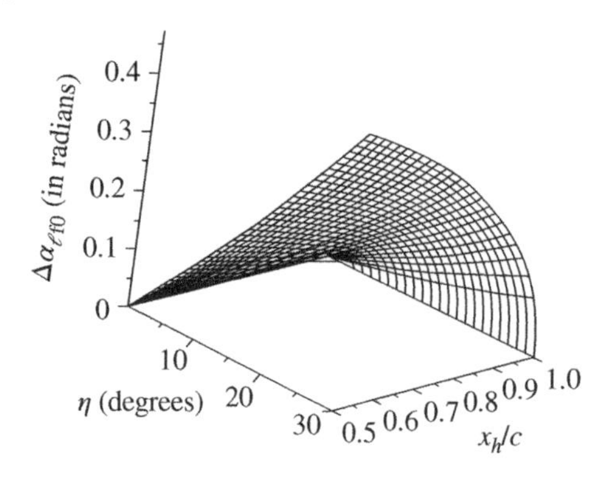


**Figure 5.13** $\Delta\alpha_{\ell f0}$ (in radians) as a function of $x_h/c$ and $\eta$ ranging from 0 to 30°

## 5.8 Enrichment Topics

### 5.8.1 Thickness Contribution to the Lift

Part (c) in Figure 5.1 depicts airfoil thickness and thickness distribution, and the question is how big is this contribution compared to other equations in this chapter? A convenient relation is given in Eq. (5.24), one that has traditionally been used to estimate the effect of *thickness ratio* $(t_m/c)$ on the lift slope $c_{\ell\alpha}$. Here, $t_m$ is the maximum airfoil thickness and $c$ is the chord length. The lift coefficient increase on early NACA airfoil designs is given as

$$\Delta c_{\ell a} = 2\pi\left(\frac{1 + 1.54(t_m/c)}{1 + 0.77(t_m/c)}\right) \tag{5.24}$$

Many common thin airfoils have $(t_m/c)$-ratios below 15% so that increases in the slope of the lift coefficient due to thickness can be expected to add less than 10% to $c_{\ell\alpha}$. Because of the presence of boundary layers, data on the effect of airfoil thickness do not always reflect any increase in $\Delta c_{\ell a}$ with different values of $t_m/c$. Thickness distribution designs, however, are necessarily aimed at boundary layer management in order to avoid separation at angles of attack greater than 15°. Newer sectional designs like the NASA LS(1)-04xx do better than Eq. (5.24) predicts.

The importance that thickness ratio $(t_m/c)$ and thickness distribution play in airfoil performance is primarily seen when the boundary layers are not being properly controlled. Because the boundary-layer behavior is highly complicated, there have been no reliable general guidelines on the effects of thickness.

### 5.8.2 Longitudinal stability of a wing

The $x$-location of the center of gravity (*cg*) – the $x$-location where its total mass vector acts – with respect to the wing's aerodynamic center (*ac*) is critical for an airplane's behavior during unsteady flight. Being the chord location where moment coefficients are constant, the aerodynamic center is a particularly important reference point in a wing's longitudinal (i.e. along the direction of travel) stability during flight. Consider a lift increase due to a sudden wind gust; if the aerodynamic center (*ac*) is behind *the center of gravity* (*cg*) of the wing, then this increase in lift creates a "nose-down" moment about the center of gravity which tends to return the airfoil to its equilibrium orientation. A *destabilizing effect* takes place, however, when the aerodynamic center is located *ahead of the center of gravity*. We note that the aerodynamic center is relatively independent of camber and thickness, and in subsonic flows it is located at ¼-chord in contrast to supersonic flows where it is found at ½-chord.

An example of the above-mentioned stability problems arises when an existing aircraft is fitted with heavier engines than originally designed for because this can move the center of gravity but not the aerodynamic center since the wings and fuselage remain unchanged.

## 5.9 Summary

The concept of a vortex sheet has been introduced in this chapter to represent the aerodynamics of thin airfoil sections in incompressible flows. This vortex sheet models the airfoil as a fixed slip-line in ideal flows that may have a variety of appropriately small camber-line profiles that also include flat plates and wings with flaps. The resulting lift coefficient for a flat plate at angles of attack below 20° is simply $c_\ell = 2\pi\alpha$, and camber introduces a zero-lift angle contribution (where $\alpha_{\ell 0} < 0$) to this equation as shown here

$$\boxed{c_\ell = 2\pi(\alpha - \alpha_{\ell 0})} \tag{5.17}$$

With both camber and flap deflections are added to a flat plate at angle of attack, we arrive at the following set of equations, where $\theta_h$ is the $x$-location of the flap-hinge in transformed coordinates,

$$c_\ell = 2\pi\alpha + 2\int_0^\pi \left(\frac{dz}{dx}\right)_w (\cos\theta - 1)d\theta + 2[(\pi - \theta_h) + \sin\theta_h]\eta \tag{5.25}$$

$$c_{mc/4} = \frac{1}{2}\int_0^\pi \left(\frac{dz}{dx}\right)_w (\cos 2\theta - \cos\theta)d\theta + 1/2[\sin\theta_h(\cos\theta_h - 1)]\eta \tag{5.26}$$

$$\alpha_{\ell 0} = -\frac{1}{\pi}\int_0^\pi \left(\frac{dz}{dx}\right)_w (\cos\theta - 1)d\theta - \frac{1}{\pi}[(\pi - \theta_h) + \sin\theta_h] \tag{5.27}$$

Equation (5.25) clearly presents results from our thin airfoil theory basic assumption, namely, that at moderately small angles of attack, camber and flap deflections each contribute linearly to the lift coefficient. The net effect of camber is to shift the $c_\ell$ vs $\alpha$ curve by an amount $\alpha_{\ell 0}$ and to increase $c_{\ell max}$ without changing the lift-slope or the functionality of the initial curve.

As is shown in Chapter 11, using sin($\alpha$) instead of the angle $\alpha$ itself in Eqs. (5.10a) and (5.25) is more accurate and permits the use of angles in degrees instead of radians which is what most trigonometric tables and electronic calculators' default to.

## Problems

**5.1** A two-dimensional thin airfoil under ideal flow has a zero-lift angle of attack of – 1.8°.
a) Calculate $c_\ell$ for $\alpha = 5°$.
b) Sketch $c_\ell$ vs $\alpha$ when the region before airfoil stall is $\alpha = \pm 12°$.
c) Is the moment about the quarter-chord location positive, negative, or zero?

**5.2** Using Eq. (E5.1) from Example 5.1, calculate $\Delta C_p$ for a symmetric airfoil and compare with data shown in Figure 5.6. Find the $-\Delta C_p$ values by subtracting the information in the graph at each $x/c$ value. Take $x/c = 0.1$, 0.3, 0.5, and 0.7.

**5.3** The $C_p$-distribution shown has been "stylized from data" for a certain two-dimensional airfoil at angle of attack. Calculate the lift coefficient per unit span.

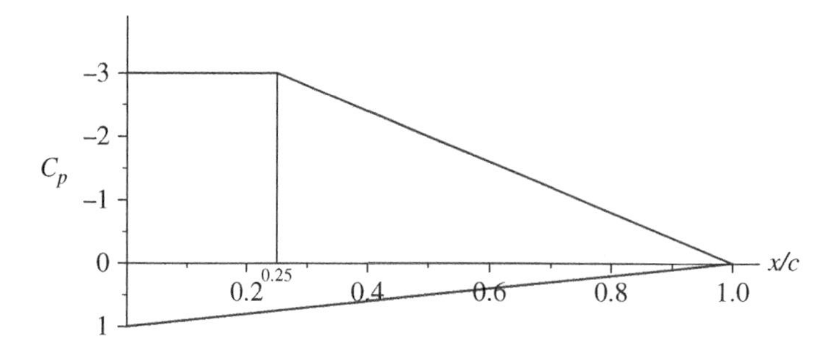


**Figure P5.3** $C_p$ vs $x/c$.

**5.4** Pressure distribution measurements around an airfoil section may be given in terms of simplified curve fit for the pressure coefficient. Data for $\alpha = 18°$ on a certain two-dimensional airfoil are expressed below. Calculate the corresponding value for the lift coefficient $c_\ell$.

$$\text{Upper surface:}\ C_{pu} = \frac{-0.7}{x/c + 0.1} + 0.7$$

$$\text{Lower surface:}\ C_{pl} = 1 - (x/c) \quad 0 \quad \leq x/c \leq \ 1.0$$

**5.5** Find the expression for the sectional moment coefficient about the aerodynamic center for a two-dimensional thin airfoil with a parabolic mean-camber line of $(dz/dx)_w = (4z_m/c)\cos(\theta)$, $0 \leq \theta \leq \pi$.

**5.6** A thin, two-dimensional cambered airfoil is made of two straight line segments as detailed below. Calculate $c_\ell$ and $c_{mac}$ for $\alpha = 0°$.

$$(dz/dx)_w = 0.08 \quad 0 \leq x/c \leq 0.5 \text{ and } (dz/dx)_w = -0.08 \quad 0.5 \ \leq x/c \leq \ 1.0$$

**5.7** A thin, two-dimensional symmetric airfoil has a plain flap which is deflected 15°. When the geometric angle of attack is $\alpha = 6°$, what are the values of the total section lift coefficient $c_\ell$ and the moment coefficient about the aerodynamic center $c_{mc/4}$? The flap length is $0.18c$ where $c$ is chord.

**5.8** Using Figures 5.8 and 5.11, find the total sectional lift coefficient for an NACA 64-618 airfoil at $\alpha = 4°$ with an extended flap at $x_h/c = 0.8$ and $\eta = 17°$. Assume additive contributions.

**5.9** Pick one of the three NACA airfoils whose camber line equations are given below. Using Eqs. (5.18) and (5.19), calculate the resulting angle of attack at zero lift ($\alpha_{\ell 0}$) and $c_{mc/4}$. Compare your results with data on these NACA airfoils.

| | |
|---|---|
| NACA 2412 – four-digit airfoil | |
| $z/c = 0.125[0.8(x/c) - (x/c)^2]$ | $0 \leq x/c \leq 0.4$ |
| $= 0.0555[0.2 + 0.8(x/c) - (x/c)^2]$ | $0.4 \leq x/c \leq 1.0$ |
| NACA 4412 – four digit airfoil | |
| $z/c = 0.25[0.8(x/c) - (x/c)^2]$ | $0 \leq x/c \leq 0.4$ |
| $= 0.111[0.2 + 0.8(x/c) - (x/c)^2]$ | $0.4 \leq x/c \leq 1.0$ |
| NACA 23012 – five-digit airfoil | |
| $z/c = 2.6595[(x/c)^3 - 0.6075(x/c)^2 + 0.11471(x/c)]$ | $0 \leq x/c \leq 0.2025$ |
| $= 0.02283(1 - x/c)$ | $0.2025 \leq x/c \leq 1.0$ |

## Check Test

**5.1** What are the three contributions to the lift in thin airfoils theory? Are they all of similar magnitude?

**5.2** What are the units of $\gamma$ and $\Gamma$ and how are these two vorticity variables related? Give your answer in the SI system.

**5.3** What is the effect of camber on the lift coefficient? Does it change its slope in the linear region?

**5.4** A rotating cylinder in a uniform stream develops a circulation $\Gamma$. What would the angle of attack on a symmetric airfoil of chord length "$c$" have to be to develop the same lift per unit span in the same stream?

**5.5** What restriction does the Kutta condition impose on $\gamma(x)$ the vortex filaments spread in a camber line?

**5.6** Show that $\Delta p = \rho V_\infty \gamma$ or equivalently that $C_p = -2\gamma V_\infty$, where $\gamma$ is the strength or circulation density of a vortex sheet, $\Delta p$ is the pressure difference between the upper and lower surfaces, $\rho$ is the gas density, and $V_\infty$ is the free stream velocity. Is this result restricted to symmetrical airfoils?

# 6

# Thin Wings of Finite Span in Incompressible Flow

## 6.1 Introduction

In this chapter, we complete our treatment of thin wings in incompressible flows under ordinary lift conditions. In Chapter 5, we examined how a two-dimensional airfoil section derives lift from the interaction of its bound circulation with the free stream velocity. Such situation represents only sectional characteristics or wings of infinite span. Real wings, however, have finite spans and the lift over these wings must be so distributed that it vanishes at the wing tips where lower pressures from above the wing meet higher pressures coming from below. Amounts of wing circulation must therefore vary along the span according to the distribution of lift, ordinarily peaking in the middle and vanishing at the tips. It turns out that such spanwise lift variations generate a significant retarding force called *drag due to lift* or *induced drag* adding to the skin drag from the boundary layers.

In this chapter, we conclude our discussion of classical aspects of finite-wing theory wherein the *potential vortex* introduced in Chapter 3 has been treated as a portion of a *vortex filament*. We may represent any required spanwise *vortex line distribution* with a large group of these filaments judiciously located along the wing. Since ideal-flow vortex filaments can only exist as closed loops, this model leads to the notion of end loops of vorticity emerging from the wing. In time, as the vortex filament attached to the wing moves away from its starting location, its stationary back end has a negligible influence on the flow over the wing and this has given rise to the concept of a *horseshoe vortex*. Moreover, since along each filament vortex strength must be constant, requisite variations of span circulation are achieved by adding vortex filaments of different widths along a location on the wing called "the lift line." In this model, we superpose vortex strengths along the wing's span in contrast to Chapter 4 where we added filaments along the chord. Under certain meteorological conditions, vortices shed from wing tips are observable in the wakes of flying airplanes and this confirms that some form of embedded vortex flow must be present on a lifting wing.

## 6.2 Objectives

Successful completion of this chapter will be evidenced by your ability to:

1) Describe the reason for trailing vortices issuing from on a wing.
2) Discuss the concept of the *horseshoe vortex* and its role in the lifting line theory.
3) Define *downwash velocity* and how it induces drag.
4) Draw a vector diagram showing the downwash angle, together with the lift and induced drag at an arbitrary span location along a finite wing.

*Elements of Aerodynamics: A Concise Introduction to Physical Concepts*, First Edition. Oscar Biblarz.

Companion website: www.wiley.com/go/elementsofaerodynamics

5) Write down the elliptic lift distribution equation along with its resulting equations for lift, induced drag, and downwash angle.
6) Describe how elliptic lift distributions can be produced.
7) Explain how the "parabolic drag polar" for an elliptic planform is constructed.
8) For the elliptic distribution, show the effect of aspect ratio on the slope of the $C_L$ vs $\alpha$ curve, i.e. on $C_{L,\,\alpha}$.
9) Explain in your own words the meaning of Eqs. (6.12) and (6.17).

## 6.3 Lifting Line Theory

As pictured in Figure 6.1, a bound vortex line that models a finite-length airfoil has to bend at the wing tips. This is shown as an open or "horseshoe vortex" and not as a vortex loop because the closing end is considered to be far enough behind so as to have no influence on the lifting line. Two pioneers in the field, Prandtl and Lanchester, independently proposed this ingenious model, one that successfully explains the most important features of finite wings in incompressible flows.

In this chapter, we will concentrate on the so-called "elliptical distribution of lift" because of its theoretical importance. Discussion of other circulation spanwise distributions and their associated equations is relegated to Appendix B. Since we are dealing with three dimensions, $y$ is the coordinate along the span and $z$ is in the lift direction.

## 6.4 Downwash Velocity and Elliptic Spanwise Lift Distribution

To generate the needed lift force profiles, a collection of bound distributed vortices is required, and each horseshoe vortex component in the set induces a downward flow velocity at the lift line. This velocity is called the *downwash* ($w$) and arises from the trailing portion of each horseshoe vortex. In order to formulate the resulting effects of this downwash, we must recognize, as seen in Figures 6.2 and 6.3, that such downward velocity introduces a new angle of attack ($\alpha_i$) that modifies the original free stream flow direction. Figure 6.2, reproduced from Kuethe and Chow (1998), captures all the significant detail that will be dealt with in this chapter. In this figure, the product "$\rho V\Gamma$" represents both the vertical lift and induced drag vectors and is drawn perpendicular to the direction of $V_{eff}$

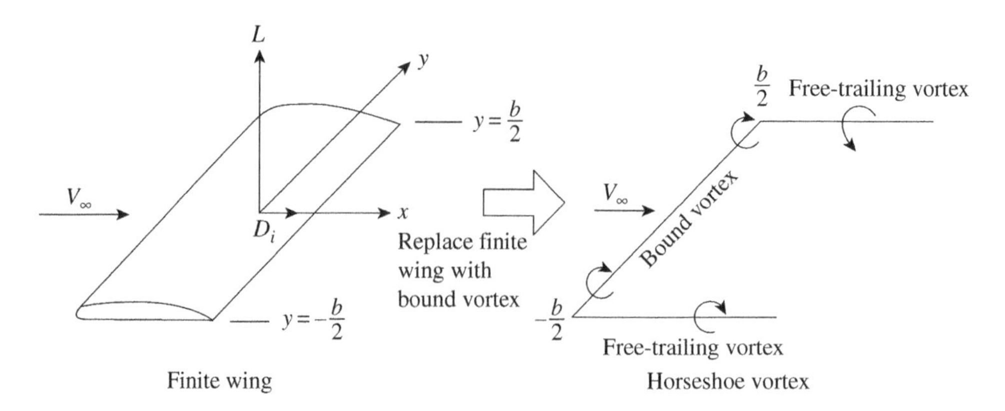


**Figure 6.1** Bound vortex substituting for a finite wing. *Source:* From Anderson (2017).

which is more clearly depicted in Figure 6.3. Our notation differs from Figure 6.2 in that $L'$ is our $\ell$ and $Di'$ is our induced drag per unit span $d_i$. Figure 6.2 also shows much detail that relates to contributions at the lifting line of the trailing vortices under summation (these are identified by their angular coordinates $\beta$ and $r$ on a trailing vortex element $dx$). The sum of individual trailing vortices will become an integral when the horseshoe vortex distribution is merged into a continuous sheet.

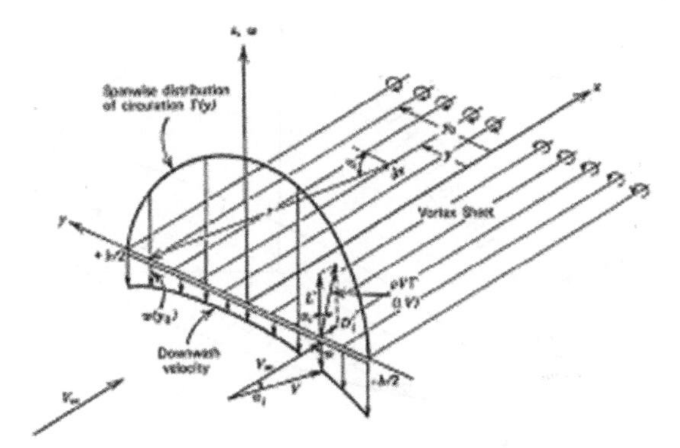


**Figure 6.2** Finite-wing lift distribution depicting the collection of discrete horseshoe vortices that generate it. *Source:* From Kuethe and Chow (1998).

At any given span location "$y$," we write the resulting lift ($\ell(y)$) and induced drag ($d_i(y)$) using the Kutta–Joukowski theorem in terms of the velocity vector $V_{eff} = V_\infty \cos\alpha_i$ which together with the relation for the downwash velocity become

$$\ell(y) = \rho_\infty V_\infty \Gamma(y) \cos\alpha_i \text{ and } d_i(y) = \rho_\infty V_\infty \Gamma(y) \sin\alpha_i \tag{6.1}$$

$$w(y) = V_\infty \tan\alpha_i(y) = V_\infty \alpha_i(y) \tag{6.2}$$

Since $\alpha_i$ is a very small angle, we have approximated the trigonometric functions as

$$\sin\alpha_i = \tan\alpha_i = \alpha_i \text{ and } \cos\alpha_i = 1.0$$

Showing the induced or downwash velocity explicitly, we arrive at the following:

$$\ell(y) = \rho_\infty V_\infty \Gamma(y) \quad \text{and} \quad d_i(y) = \rho_\infty V_\infty \Gamma(y)\alpha_i(y) \equiv \rho_\infty \Gamma(y) w(y) \tag{6.3}$$

Calculations of induced drag with this model involve adding induced velocities from each trailing component of the horseshoe vortex sheet along the span. This model turns out to provide good estimates of observed behavior. We shall omit the mathematical detail and go directly to its results based on the downwash velocity $w(y)$.

Prandtl found that for an *elliptical $\Gamma(y)$-distribution,* the downwash becomes constant along the span and the induced drag a minimum, see Figure 6.4, whereas Figure 6.2 shows the downwash from an arbitrary distribution. Moreover, it can be shown that elliptical results are generated by the elliptical-shaped planform in Figure 6.4 and quite closely by many of the more practical trapezoidal ones. For arbitrary planform shapes, the calculation of $\alpha_i$ is not particularly straightforward (e.g. see Anderson 2017; Bertin and Cummings 2013; Kuethe and Chow 1998; Schlichting and Truckenbrodt 1979 among others); additional information is given in Appendices B and C.

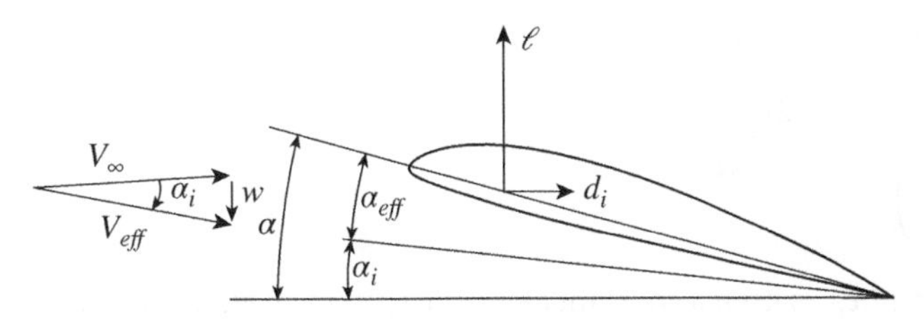


**Figure 6.3** Airfoil section showing consequences from its finiteness. $V_{eff}$ represents the "effective" velocity vector affecting the three-dimensional section. The bending of $V_\infty$ splits the geometric angle of attack into $\alpha = \alpha_i + \alpha_{eff}$, where $\alpha_{eff}$ represents the modified or effective flow direction that the airfoil experiences.

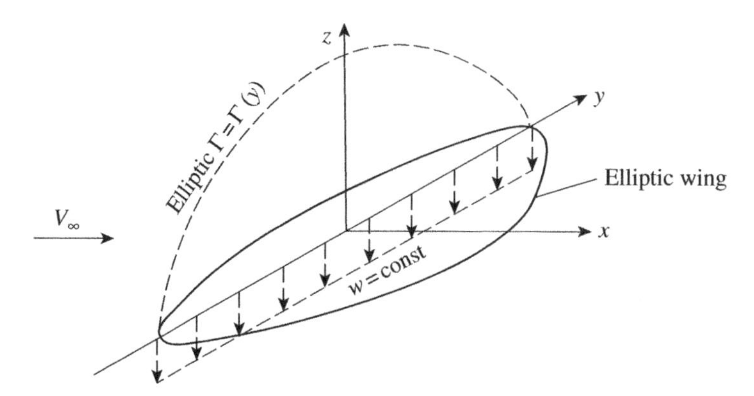


**Figure 6.4** Elliptical planforms that generate elliptical lift distributions. *Source:* From Anderson (2017).

For a wing during cruise mode, the total elliptical $\Gamma$-distribution is symmetric about $y = 0$, at the span center; and if we let $\Gamma(0) = \Gamma_s$, this circulation distribution may be written as (where $b$ is the total span)

$$\Gamma(y) = \Gamma_s \sqrt{1 - \left(\frac{y}{b/2}\right)^2} \tag{6.4}$$

As depicted in Figure 6.2, each individual trailing vortex contributes to the downwash at any given lift-line location $y_0$ and all these must be summed. The following relation shows the resulting formulation for the induced angle $\alpha_i$ from the vortex sheet components.

$$\alpha_i(y_0) = \frac{1}{4\pi V_\infty} \int_{-b/2}^{b/2} \frac{d\Gamma/dy}{y_0 - y} dy \tag{6.5}$$

The result of this integration, after properly substituting equations (6.4) into (6.5), shows that the elliptical distribution's downwash angle is not only the same at all $y_0$-locations but also the minimum possible value of $\alpha_i$ and, according to Eq. (6.2), of the downwash velocity. Summarizing, for an elliptic distribution generated with an elliptic planform, we arrive at the simple but important result [check the units of the components of this equation!]

$$w(y_0) = \alpha_i V_\infty = -\Gamma_s/2b \text{ (constant along the span)} \tag{6.6}$$

We proceed now to calculate the aerodynamic coefficients. However, since we are not dealing with a rectangular planform, we first need to establish the following relations for the *elliptical chord* $c_e(y)$ and the *elliptical sectional lift coefficient* $c_{\ell e}$ (where $c_{root}$ is the value of the chord at $y = 0$):

$$c_e(y) = c_{root} \sqrt{1 - \left(\frac{y}{b/2}\right)^2} \tag{6.7}$$

$$c_{\ell e} = \frac{\rho_\infty V_\infty \Gamma_s \sqrt{1 - \left(\frac{y}{b/2}\right)^2}}{1/2 \rho_\infty V_\infty^2 c_{root} \sqrt{1 - \left(\frac{y}{b/2}\right)^2}} = \frac{2\Gamma_s}{V_\infty c_{root}} \tag{6.8}$$

From this we can see that, similar to $w$, $c_{\ell e}$ is itself a constant along the span of an elliptical wing. We next integrate as we did back in Eq. (5.9) but here over $y$ to get the total airfoil lift, observing that the elliptical planform area ($S_e$) is given by

$S_e = (\pi/4)bc_{root}$

Incidentally, this surface area amounts to about 80% of a rectangular planform for which $c_{root}$ is the same everywhere. Applying next the sectional lift from leftmost portion of Eq. (6.3) to calculate the total lift ($L_e$) and its corresponding coefficient ($C_{Le}$),

$$L_e = \rho_\infty V_\infty \Gamma_s \int_{-b/2}^{b/2} \sqrt{1 - \left(\frac{y}{b/2}\right)^2} dy = (\pi/4)\rho_\infty V_\infty \Gamma_s b \tag{6.9}$$

$$C_{Le} = \frac{L_e}{1/2\rho_\infty V_\infty^2 S_e} = \frac{(\pi/4)\rho_\infty V_\infty \Gamma_s b}{1/2\rho_\infty V_\infty^2 (\pi/4) b c_{root}} = \frac{2\Gamma_s}{V_\infty c_{root}} \equiv c_{\ell e} \tag{6.10a}$$

$$C_{Le} = \frac{(\pi/4)\rho_\infty V_\infty \Gamma_s b^2}{1/2\rho_\infty V_\infty^2 S_e b} = \frac{\pi \Gamma_s \mathbf{AR}}{2bV_\infty} \tag{6.10b}$$

Equation (6.10a) demonstrates that for elliptical planforms, the total lift and the sectional lift coefficients are equal to each other. Eq. (6.10b) shows the form of the total lift coefficient in terms of the wing's aspect ratio, ***AR***.

Now the total induced drag ($D_{ei}$) and its slope ($C_{Dei}$) for an elliptical wing, using the second part of Eq. (6.3) along with Eq. (6.5), become

$$D_{ei} = \rho_\infty V_\infty \int_{-\frac{b}{2}}^{\frac{b}{2}} \Gamma_s \sqrt{1 - \left(\frac{y}{b/2}\right)^2} \left(\frac{-\Gamma_s}{2bV_\infty}\right) dy = \frac{\pi}{8} \rho_\infty V_\infty b \Gamma_s^2 \tag{6.11}$$

$$C_{Dei} = \frac{D_{ei}}{1/2\rho_\infty V_\infty^2 S_e} = \frac{\pi b \Gamma_s^2}{4V_\infty S_e} = \frac{C_{Le}^2}{\pi}\left(\frac{S_e}{b^2}\right) = \frac{C_{Le}^2}{\pi \mathbf{AR}} \tag{6.12}$$

In elliptical wings with components $C_{Le}$ and ***AR***, the induced drag for these planforms as given in Eq. (6.12) is a minimum. For other wing planforms, we introduce *e* the *span efficiency factor* ($e < 1.0$ for non-elliptic planforms)

$$\boxed{C_{Di} = C_L^2/(\pi e \mathbf{AR})} \tag{6.13}$$

Including skin drag ($c_d$), a total drag coefficient for incompressible flows over a wing then becomes

$$C_D = c_d + C_L^2/(\pi e \mathbf{AR}) \tag{6.14}$$

For many common wing designs, the downwash velocity along the span remains close to constant. Practical elliptical lift distributions are found in the untwisted wings of many trapezoidal planforms so their absolute angle of attack has nearly the same value everywhere along the span. Since their geometric angle of attack and downwash angle do not much differ, it follows that effective angle of attack and local lift coefficient may also be considered nearly constant along the span and close to the value from elliptical lift distributions.

Using our results for the symmetric portion of the *elliptic planform* airfoil, we now proceed to calculate a useful relation between the lift coefficient ($C_{Le}$) and angle of attack $\alpha$ for finite wings equivalent to Eq. (5.10b). From Eqs. (6.6) and (6.10b),

$$\alpha_i = -\frac{\Gamma_S}{2bV_\infty} = -\frac{C_{Le}}{\pi \boldsymbol{AR}}$$

Using Eq. (6.10a) and interpreting the vectors displayed in Figure 6.3,

$$c_{\ell e} = 2\pi(\alpha_{eff}) = 2\pi\left(\alpha - \frac{C_{Le}}{\pi \boldsymbol{AR}}\right) \equiv C_{Le} \tag{6.15}$$

$$C_{Le}\left(1 + \frac{2\pi}{\pi \boldsymbol{AR}}\right) = 2\pi\alpha$$

$$C_{Le} = \frac{2\pi\alpha}{\left(1 + \frac{2}{AR}\right)} \tag{6.16}$$

The slope of $C_{L\alpha}$ in Eq. (6.16) may be generalized for wings other than elliptic by introducing $a_0$ as a two-dimensional slope and $a$ as its three-dimensional counterpart, and inserting $e$ as the span efficiency factor ($e$ <1.0 and $a_0 \leq 2\pi$),

$$\boxed{C_{L\alpha} \equiv a = \frac{a_0}{\left(1 + \frac{a_0}{\pi e \boldsymbol{AR}}\right)}} \tag{6.17}$$

As we did in Chapter 5, the cambered portion of the airfoil at zero angle of attack shown in Figure 5.1 may be added to the symmetric portion and that entails the introduction of the zero-lift angle of attack ($\alpha_{\ell 0}$) into Eq. (6.16). The generalized version is shown as Eq. (6.22) in Section 6.9.

**Example 6.1** Using the nearly elliptical planform results for the total drag coefficient in Eq. (6.14) and the three-dimensional slope in Eq. (6.17), shows that these are consistent with results in Chapter 5 as the ***AR*** goes to infinity (i.e. they become our two-dimensional results).

Consider a nearly rectangular wing of constant chord $c$ and variable span $b$. Elliptic and/or other wing planform shapes may be related to such wing through a corresponding sizing factor.

$$\boldsymbol{AR} = b^2/S \approx b^2/bc = b/c \quad \text{as} \quad b \rightarrow \infty \quad \text{then} \quad C_{Di} = C_L{}^2/\pi e \boldsymbol{AR} \rightarrow (L/q_\infty)^2/\pi e b^3 c \rightarrow 0$$

Now $C_L = a\alpha$, where $C_L = L/(q_\infty bc) \rightarrow c_\ell$ and $a \leq 2\pi/(1+2\pi/\pi e \boldsymbol{AR}) \rightarrow a_0$

$$C_D = c_d + C_L{}^2/\pi e \boldsymbol{AR} \rightarrow c_d \quad \text{and} \quad a \rightarrow a_0 \leq 2\pi\alpha$$

In the real world, very wide wings are impractical; the larger values of ***AR*** would result in heavier wings because more structural components are needed. In addition, more difficult ground maneuvering for the aircraft and greater manufacturing expenses become undesirable factors.

## 6.5 Experimental Verification Using Drag Polars

A *drag polar* representation combines in one graph information for the total lift and total drag coefficients as shown in Figure 6.5. It is essentially a plot of Eq. (6.14) from measurements. When experimental values fall within the linear $C_L$ vs $\alpha$ region and plot as a parabola in the polar version, the

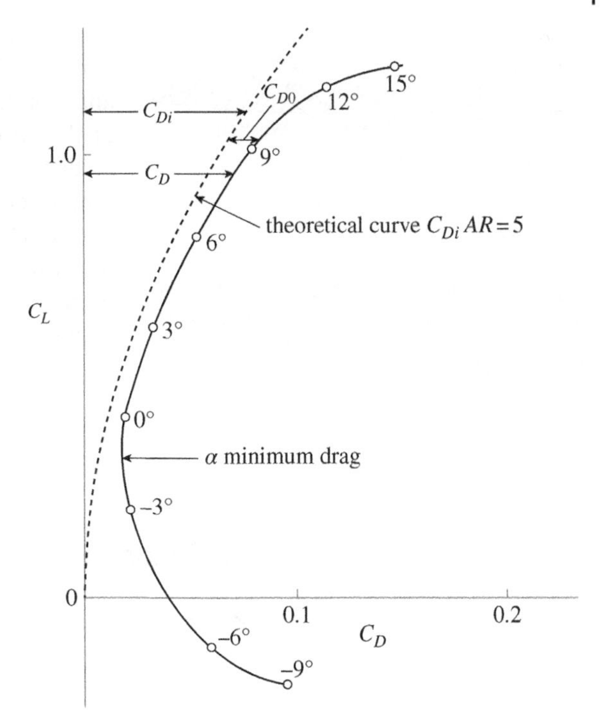


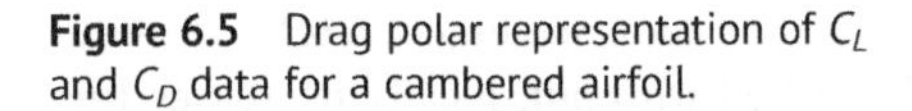
**Figure 6.5** Drag polar representation of $C_L$ and $C_D$ data for a cambered airfoil.

constant offset in Eq. (6.14) amounts to the viscous drag component ($C_{D0}$ or $c_d$), and the model given in this chapter accurately represents observations. This plot can also be shown to provide good rapid estimates of real behavior for wings with various ***ARs*** (as originally done by Prandtl). Small corrections may be brought in empirically for non-elliptical conditions. In Figure 6.5, the $\alpha$-location for minimum drag is identified as the leftmost portion of the solid curve. Regions of nonlinear $C_L$ vs $\alpha$ data may be inferred by the lack of similarity between the solid curve and the dashed parabolic one and from the angle of attack information that appears in this figure.

## 6.6 Non-elliptic Planforms and Twist

Elliptic wing planforms can minimize induced drag but are relatively expensive to manufacture, and aircraft with elliptic wings are difficult to maneuver in flight. It turns out, moreover, that many ordinary incompressible-flow wing designs perform only slightly less efficiently than the elliptic – about 10% for trapezoidal wings. Thus, there has been a practical incentive to utilize non-elliptic planforms in small civilian aircraft. Appendix C gives details on the conventional Fourier series analysis for arbitrary circulation distributions, and and Appendix B lists some available software programs that perform such calculations.

*Wing twist* is a technique known to reduce a portion of the extra induced drag of non-elliptic planforms. Wing *twist* is also a good way to alleviate a *stall-defect* of tapered planforms at high angles of attack. In order for ailerons (moveable control surfaces located at the wing's rear edges) to remain effective, stall patterns should not develop over their surfaces. In *geometric twist*, the outer portions of the airfoil physically rotate upward relative to the root portions to attain a continuously

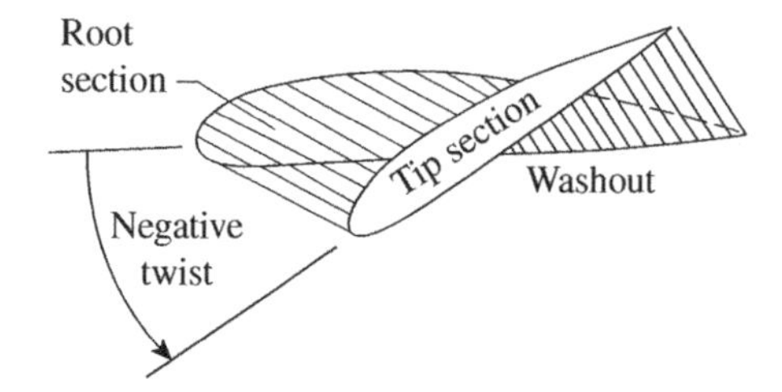


**Figure 6.6** Geometric twist at a wing tip. For such wings, the angle of attack becomes $\alpha(y)$ because it varies along the span. *Source:* From NASA SP-367.

decreasing angle of attack, see Figure 6.6. Such twist, which is also called *washout*, locally decreases the angle of attack to below the stall angle of the rest of the wing, and this may require the tip airfoil section to have a lower degree of camber than the root section that results in "negative twist" as in Figure 6.6. However, since the degree of twist for best overall results depends on variables such as $q_\infty$and $\alpha$, it is more desirable to manufacture wings with adjustable amounts of twist; there is some expectation that such *morphing wings* will be soon achievable by combining artificial intelligence (AI) with new materials and manufacturing methods – some details are given in Chapter 11.

## 6.7 Effects of Lifting Line Theory on Airplane Performance

### 6.7.1 Optimum Cruising Speed (Maximum *L/D*)

As introduced, an efficiency factor ($e$) is used in Eq. (6.17) to represent real wings where $e \approx 0.98$ or an entire airplane where $e \approx 0.6$ to 0.9. From Eq. (6.14), the total drag dependence on the cruising speed $V_\infty$ becomes (here $k_1$ and $k_2$ are constants)

$$D = c_d \frac{\rho V_\infty^2 c}{2} + \frac{4L^2}{\rho^2 V_\infty^4 S\pi e\boldsymbol{AR}} \frac{\rho V_\infty^2 c}{2} = k_1 V_\infty^2 + \frac{k_2}{V_\infty^2}$$

That is, frictional drag increases proportionally to $V_\infty^2$ and induced drag decreases proportionally to $1/V_\infty^2$. It is therefore expected that the ratio of $D/L$ minimizes or that an *optimum* speed exists as shown below. Using our definitions for the lift and drag coefficients in Chapter 2 together with Eq. (6.14), we have

$$\frac{D}{L} \equiv \frac{C_D}{C_L} = \frac{c_d}{C_L} + \frac{C_L}{\pi e\boldsymbol{AR}} \tag{6.18}$$

which becomes a minimum when drag due to lift equals all other drag contributions, which are called here *parasite drag*. This minimum condition is shown in Figure 6.7.

$$C_{Di} = c_d = C_L{}^2/(\pi e\mathbf{AR})$$

so that there the total drag may be written as

$$C_D = 2c_d$$

Finally, solving the inverse of Eq. (6.18) and $C_L$ at the maximum lift-to drag location, we obtain

$$\left(\frac{L}{D}\right)_{max} = \sqrt{\frac{\pi e\boldsymbol{AR}}{4c_d}} \tag{6.19a}$$

$$C_L = \sqrt{\pi e\boldsymbol{AR}c_d} \ \text{ at } \ (L/D)_{max} \tag{6.19b}$$

*Parasite drag* is a name often given to the portion of the total airfoil drag coefficient that does not depend on lift. In Chapter 7, we examine this drag more fully, but here we simply identify it as skin

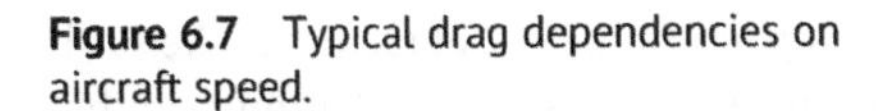

**Figure 6.7** Typical drag dependencies on aircraft speed.

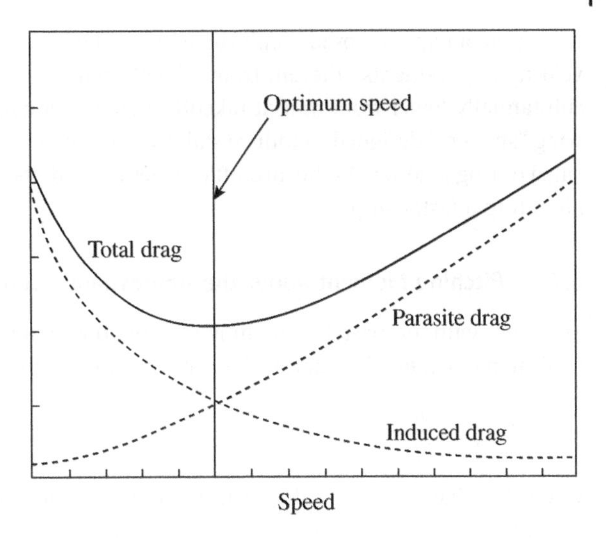


friction and pressure drag and retain the two-dimensional symbol $c_d$ since any spanwise drag variations are expected to be negligibly small.

**Example 6.2** An airplane is designed to travel at 150 km/h where its $L/D)_{max} = 12$. It has a wing planform area of 24 $m^2$ with a span of 12 m and an efficiency factor of $e = 0.9$. It weighs 21 600 N and cruises near sea level. Calculate the corresponding parasite drag coefficient.

At this low altitude (see Appendix A), the atmospheric density is 1.227 $kg/m^3$ and the dynamic pressure becomes $q_\infty = 1065$ $N/m^2$. The aspect ratio for this wing is $\boldsymbol{AR} = (12)^2/24 = 6.0$. Using $L = W$ $C_L = W/q_\infty S = 0.846$ and the induced drag $C_{Di} = C_L{}^2/\pi e\boldsymbol{AR} = 0.0422$ (5% of lift)

The total drag can be found as $C_D = C_L/12 = 0.846/12 = 0.0705$

The drag other than $C_{Di}$ becomes $c_d = 0.0705 - 0.0422 = 0.023$ or $D_0 = 587.9$ N.

In this example, the drag due to friction and other causes is 60% of the total. Note that we have not identified the other contributions to the airplane drag, but they can be shown to amount to less than about 15% of the total for well-designed aircraft.

### 6.7.2 The Stall Velocity

Next, we want to briefly look at the *stalling velocity* of an airfoil or speed at which lift can no longer support the weight of an airborne vehicle; high-lift airfoils will be discussed in more detail in Chapter 11. As evident from experimental observation, $c_\ell$ in real airfoils begins to drop abruptly as the angle of attack is increased into a nonlinear region, and such sharp drop brings significant limitations to aircraft performance. Denoting the corresponding maximum finite-wing lift coefficient as $C_{Lmax}$ and using $L = W$ for the cruise condition together with the definition of the lift coefficient, we arrive at an equation for the stall speed

$$V_{stall} = \sqrt{\frac{2W}{\rho_\infty S C_{Lmax}}} \tag{6.20}$$

For any given value of air density, planform area, and weight, it is desirable to take off near the lowest possible $V_{stall}$ value. All airfoils have fixed geometries but can increase their $C_{Lmax}$ through the action of flaps and other deployable high-lift devices. During landing, several factors make stall velocity requirements different from takeoff such as changes in $W$ – the weight at landing can be substantially lower from that at takeoff in conventionally fueled aircraft (i.e. nonelectric). Flight wing "spoilers" designed to induce stall by lowering $C_{Lmax}$ using deployable wing devices are activated during landing. As the aircraft touches ground, its wheel brakes are then applied to bring the aircraft to a faster stop.

### 6.7.3 Pitching Moment about the Aerodynamic Center

For wings without twist, the pitching moment does not vary along the span and we may define the total moment coefficient about the aerodynamic center as

$$C_{mc/4} = \frac{M_{a.c.}}{q_\infty Sc} \tag{6.21}$$

where $M_{a.c.}$ has the same value as its sectional counterpart ($m_{a.c.}$) in Chapter 5. As before, we take the aerodynamic center to be at $x_{a.c.} = c/4$ for wings in incompressible flows.

## 6.8 Enrichment Topics

### 6.8.1 Sail Iceboats

Sail ice-boats and sailboats in water are both propelled by the wind, but because icy surfaces produce relatively small resistive forces, ice-boats can be made to travel faster than the wind itself. Such fast mode of travel differs from moving directly *downwind* because here the sail acts as a cambered wing that generates thrust (horizontal lift) in the moving airstream. For any boat traveling downwind, the thrust force goes to zero when it reaches wind speed, whereas when the wind is coming from the side the sail acts as a wing generating a force normal to it, see Figure 6.8, and the thrust depends on the same parameters that govern the lift coefficient as in Eq. (6.10b) or (6.22) in Section 6.9. Since wind speed is no longer a limiting factor, the speed will only depend on the balance between an enhanced thrust and the resistance to motion from frictional forces with added the lift-induced drag. Angles of attack between $\alpha = 0^\circ$ and $45^\circ$ are commonly used, and these need to be adjusted often to accommodate variable wind directions. Sails and keels (i.e. their edges) intended for speed therefore exploit the same wing-design principles as aircraft because sails while being shaped by the wind become cambered airfoils.

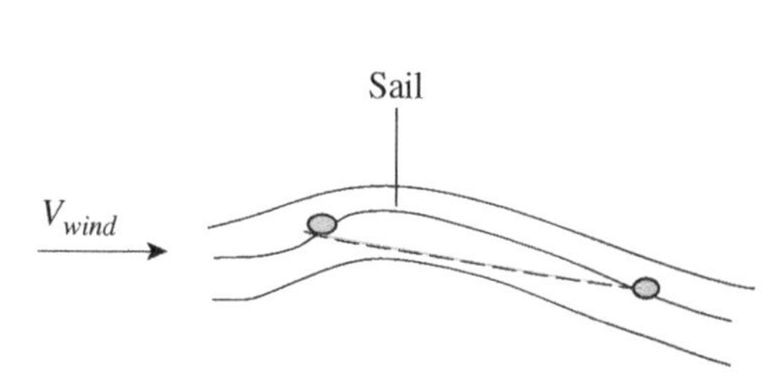


**Figure 6.8** Top view of wind flowing around a sail generating propulsive "lift."

### 6.8.2 Trailing Vortex Effects

Trailing vortices issuing from wing tips during flight give rise to several effects of significance in aerodynamics. The generation of induced drag has already been discussed in Section 6.4, so in this portion, we discuss the *ground effect* and introduce the application of *winglets* to reduce induced drag. Trailing vortices are one reason for aircraft spacings at airport runways, particularly after the takeoff of large jets, and are also relevant to the "V-flight formations" seen with migrating birds.

#### 6.8.2.1 Ground Effect

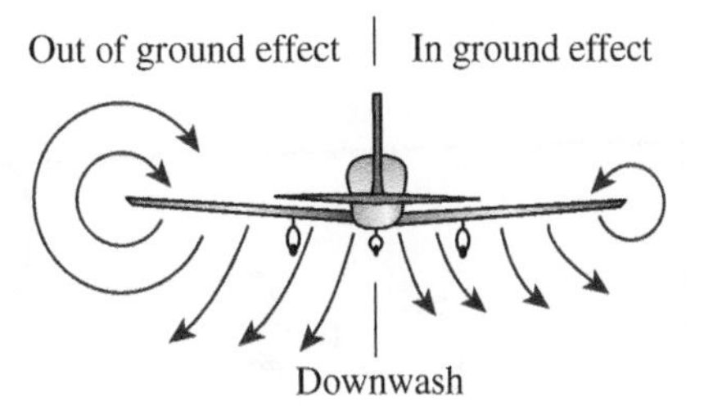


Figure 6.9 Decrease of downwash by a virtual ground-effect wing image located on the right half.

The *downwash* effect that originates from *trailing vortices* modifies large portions of the air mass above and below a moving aircraft. At or near the ground, the velocity component normal to it must vanish (i.e. no flow can be perpendicular to the solid surface) and this results on some suppression of the downwash reducing the total drag on the aircraft. This is due to a decrease of wing tip vortex strength that lessens the induced drag during takeoff and landing while leading to an effective increase in the angle of attack. This effect is also relevant to aircraft refueling during flight because account must be taken of drag and moment force changes that arise from their trailing vortex interactions between the two aircraft. Figure 6.9 depicts vortex strengths with and without ground effect during a landing.

#### 6.8.2.2 Winglets

Induced drag in subsonic aircraft is a major contributor to the total drag and much effort has been spent in designing wing attachments that minimize this drag. Because air flow over the wing span develops slight inboard components and air flow under the wing develops slight outboard components, all trailing vortices shown in Figure 6.2 eventually coalesce at the tips as they leave the wing forming two large trailing vortices.

Winglets are aerodynamically active surfaces, attached at wing tips and nearly perpendicular to the wing itself. When properly designed, they do not add significant amounts of weight and act to decrease the strength of the trailing vortices shed from the wings during flight. This in turn decreases the induced drag. But in order to be effective, winglet design must be based on local flow conditions during cruise and on specifics of the wing configuration they reside. Such designs have required sophisticated computer analyses. Figure 6.10 shows a regular wing-tip vortex shedding from the left wing and the effect much diminished by a winglet on the right wing.

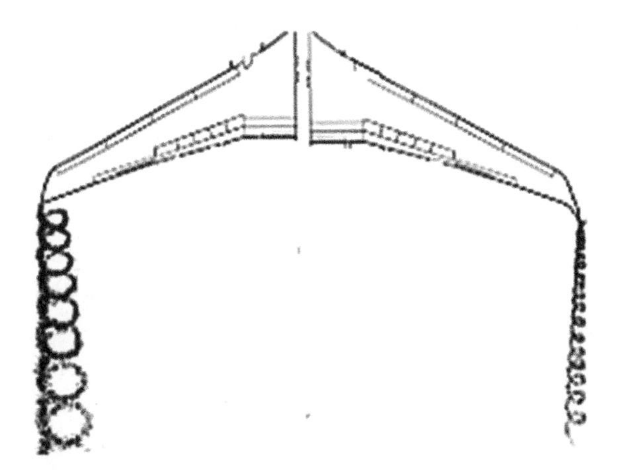

Figure 6.10 Trailing vortex profiles. The left wing shows the original shed vortex and the right wing shows a vortex shed from the same wing fitted with a winglet under the same flight conditions.

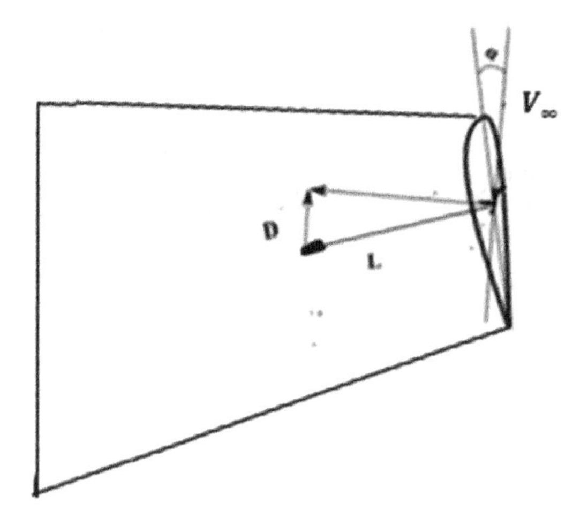


**Figure 6.11** Winglet aerodynamic setup on a right-side wing-tip.

In order to visualize forces at the wing tip, consider an airfoil section as in Figure 6.3 but rotated into a vertical plane, exchanging the $y$-axis for the $z$-axis in the figure. Flows near but outboard on finite-wing tips are nearly circular as air from below rushes outward along the span, around the tip, and inward on the upper surface. To such circular velocity, we add vectorially the much higher free stream velocity, and the resultant is a vector bent slightly inward – this situation can be visualized on a properly oriented winglet as shown in Figure 6.11, where $\phi$ is the effective angle of attack of the new total velocity vector with respect to the winglet. At the wing tips, this vector is rotated slightly in to produce a horizontal "lift force" ($L$) accompanied by some "negative drag" ($D$). Opposing "lift forces" from each tip are balanced by the wing's structure and, more significantly, the resultant negative drag subtracts from the locally induced drag. Winglets designed primarily for cruise conditions are ubiquitous in modern aircraft (as seen Figure 1.1).

## 6.9 Summary

For elliptic wings with uncambered sections in ideal flows, we obtained the following useful relations for total lift and induced drag:

$$L = \frac{\pi}{4} b\rho_\infty V_\infty \Gamma_s \tag{6.9}$$

and

$$C_L = \frac{2\Gamma_s}{V_\infty c_{root}} \tag{6.10a}$$

$$D_i = \frac{\pi}{8} \rho_\infty V_\infty b\Gamma_s^{\,2} \tag{6.11}$$

and

$$C_{Di} = \frac{\pi b\Gamma_s^{\,2}}{4V_\infty S} = \frac{C_L^2}{\pi \boldsymbol{AR}} \tag{6.12}$$

We also showed that elliptical-span lift distributions represent the most efficient distributions of $\Gamma(y)$ because the resulting constant downwash velocity leads to the minimum induced drag. Because elliptic planforms have poor stall characteristics during flight maneuvers and since rectangular or trapezoidal wings can be built to perform close to elliptic ones, all modern low-speed aircraft have non-elliptic planforms.

For non-elliptic planforms, a value of $e \leq 0.9$ (the planform efficiency factor) can be used with these equations as in Eqs. (6.12) and (6.16). The effect of camber may be accounted for with a non-zero $\alpha_{\ell 0}$ as in Chapter 5 and the equation form for $C_L$ is written below in terms of the three-dimensional slope ($a$) in Eq. (6.17) as Eq. (6.22).

$$C_{Di} = \frac{C_L^2}{\pi e \boldsymbol{AR}} \tag{6.12}$$

$$C_{L,\alpha} \equiv a = \frac{a_0}{1 + \frac{a_0}{\pi e \boldsymbol{AR}}} \qquad a_0 \leq 2\pi \tag{6.16}$$

$$C_L = a(\alpha - \alpha_{\ell 0}) \tag{6.22}$$

Software for trapezoidal and other planform designs is available from the Internet and one is identified in Appendix B – analyses on many wings with practical planforms require only few changes of the equations given in this chapter.

## Problems

**6.1** Consider an airplane that weighs 20 000 N and cruises in level flight at 200 km/h at an altitude of 3 km. The wings have a total surface area of 16.5 $m^2$ and an aspect ratio of 7.28. If the lift coefficient is a linear function of angle of attack and if the load distribution is elliptic, calculate the three-dimensional lift slope and induced drag coefficient.

**6.2** What geometric angle of attack is required from an elliptical, uncambered airfoil of aspect ratio 7.0 to produce a lift coefficient of $C_L = 0.67$?

**6.3** Two rectangular wings based on the same NACA 2415 airfoil section have different aspect ratios, namely, $\boldsymbol{AR}_1 = 5.0$ and $\boldsymbol{AR}_2 = 10.0$. For both wings, $\alpha_{\ell 0} = -2^\circ$, $c_\ell = 0.6$, and $c_d = 0.0065$ at $\alpha = 4^\circ$. Take $e = 0.9$ for both and assume $a_0$ is 93% of theoretical.

a) If the two wings have the same $c = 2$ m, what is the value of their spans?
b) Calculate the total drag coefficients at $\alpha = 4^\circ$ for 1 and 2 and compare.

**6.4** Before the advent of flaps some airfoils were built with variable wing areas without camber – called telescoping wings that retained the two-dimensional slope ($a_0$) unchanged. Given an airplane that weights 15 000 N and cruises in steady level flight, calculate the planform area for the two cases below. Take the planform to be rectangular and untwisted with a chord length of 1 m and $e = 0.95$. $\alpha = 7^\circ$ for both cases. The slope is close to elliptic for our purposes.

a) 300 km/h at 3,000m with $C_L = 0.5$.
b) 180 km/h at sea level.

**6.5** The lift-to-drag ratio of a sailplane is 32. It has a wing planform area of 9.4 $m^2$ and weights 3900 N. What is $C_D$ when the aircraft is in steady level flight at an altitude of 600 m at a speed of 72 km/h?

**6.6** You own a homemade drone airplane which has a flat, rectangular wing. The span is 10 ft and the chord is 2.5 ft but the airfoil sectional characteristics are unknown (except that it looks thin). Estimate the $L/\alpha$ when the plane flies in standard sea-level air ($p_\infty = 14.7$ psia, $T_\infty = 520\ ^\circ$R) at 150 mph.

**6.7** Show that for an untwisted *elliptical planform* $\Gamma_s = [4b\frac{V_\infty}{\boldsymbol{AR} + 2}]\alpha$ when $a_0 = 2\pi$.

## Check Test

**6.1** Calculate the ratio $C_{Di2}/C_{Di1}$ for two elliptic wings having the same $C_L$, where $\boldsymbol{AR}_1 = 3.0$ and $\boldsymbol{AR}_2 = 7.0$.

**6.2** State the factors that make $a < a_0$ in a finite wing?

**6.3** The downwash velocity results from effects arising at the wing tips. Explain and identify the major players in these effects.

**6.4** Does the induced drag increase between finite wings relatable through the ***AR*** to wing weight in airplanes under cruise conditions?

**6.5** (Optional) For a so-called *staggered biplane* design, the bound vortex on the upper wing induces an "upwash" velocity on the lower wing behind it. Draw a vector diagram similar to Figure 6.3 indicating what happens to the total velocity vector approaching the lower wing. Does such an upwash lead to negative drag? (see Prandtl and Tietjens 1934a, b, for the "Staggered Biplane" concept).

**6.6** Work problem 6.2.

# 7

# Viscous Boundary Layers

## 7.1 Introduction

Because of their low viscosity and much lower density compared to liquids, gases enable aircrafts to operate at higher speeds with much less demand from their engines than watercrafts. Over wings, air flows develop relatively thin *boundary layers* that enable the production of lift in subsonic airfoils through the *Kutta condition*. Of the many contributions to drag, we will only focus in this chapter on *skin friction* since it originates from the viscous nature of real flows. There are two distinct types of viscous flow, namely, laminar or low drag and turbulent or high drag, and transition from laminar to turbulent is unavoidable under ordinary flow conditions. Because of its many complexities, the study of viscous fluid flows around airfoils is mathematically and experimentally more challenging than that for ideal flows. For slender wings below stall, a smooth flat plate is an excellent model for the effects of skin friction, one that has been studied extensively. Most viscous effects are treated semi-empirically because even purely laminar flows are sufficiently complicated to require simplifying assumptions. This must have been one motivation for comprehensive early wind tunnel experiments in this field starting with the Wright brothers.

## 7.2 Objectives

After completing this chapter, you should be able to:

1) Discuss why thin boundary regions where viscous effects are most significant must develop in air flows over immersed objects.
2) Describe the physical differences between laminar and turbulent flows, and explain why the latter produce more flow resistance.
3) Identify the role and value of the Reynolds number, based on the chord-length coordinate $x$, in the transition from laminar to turbulent flows over a flat plate.
4) Describe the concept of the skin-friction coefficient $C_f$ and its relative magnitude between laminar and turbulent flows over flat-plate flows at comparable values of $Re_x$.
5) Explain the utility of the boundary layer thickness ($\delta$) concept and its relative magnitude between laminar and turbulent for flat-plate flows.
6) Describe the role that external pressure gradients play in boundary layer separation.

*Elements of Aerodynamics: A Concise Introduction to Physical Concepts*, First Edition. Oscar Biblarz.

Companion website: www.wiley.com/go/elementsofaerodynamics

## 7.3 The Boundary Layer Concept

A practical approach to the study of flow-viscosity effects on streamlined objects was introduced by Prandtl with his *boundary layer concept*. In this model the effects of viscosity are confined to relatively thin attached fluid layers over fully immersed objects, and regions external to these layers may be satisfactorily analyzed with potential flow theory. A basic premise of this model is that the fluid immediately adjacent to the surface is at rest relative to the surface (the no-slip condition). Hence, a variable speed region must develop in which the fluid's velocity increases from zero at the surface to the value at the free stream, and such increases depend on the airfoil's configuration and angle of attack among other things. And because of their thinness, boundary layer regions minimally modify the body's ideal-flow physical dimensions.

*Laminar flows* consist of smooth ribbon-like fluid layers with no significant cross-flow mixing, whereas fully *turbulent flows* are intensely cross-flow mixed and considerably more common (see Figure 7.1). All boundary layers begin laminar and transition to turbulent over an airfoil – the extent of the laminar portion is an important design input that depends on factors such as free stream turbulence, airfoil angle of attack, and surface roughness. Laminar boundary layers easily detach in adverse pressure gradients (i.e. where pressure increases as the velocity decreases such as the aft-section of an airfoil), so they are more difficult to maintain. In general, laminar regions require special airfoil configurations for pressure-profile management.

The concept of thin attached boundary layers on streamlined objects has been an implicit assumption for our analysis of airfoils from the start. Boundary layers, as already mentioned, are complex entities that unavoidably transition from laminar to turbulent and develop regions with a propensity to detach in adverse pressure gradients. Once a boundary layer has detached, ideal flow regions (i.e. where Bernoulli's equation applies) depart substantially from the size and shape of the immersed object and oftentimes flows become unsteady – both effects invalidate nearly all our formulations. It is necessary and important, therefore, to know where and when boundary layers detach and how to prevent or delay this occurrence because separation is also associated with sharp increases of *pressure drag* that substantially add to skin friction. All this complexity influences how we study airfoil behavior in various regions of operation and in different Reynolds number regimes. In comparing the pros and cons of laminar and turbulent flows, a designer must consider that, while laminar flows result in less drag, they easily separate, whereas turbulent flows while more lossy can remain attached longer. To delay separation, airfoils are fitted with "boundary layers trips" that turn the flow turbulent at preselected wing locations. A type of turbulence generating hardware is shown in Figure 7.2.

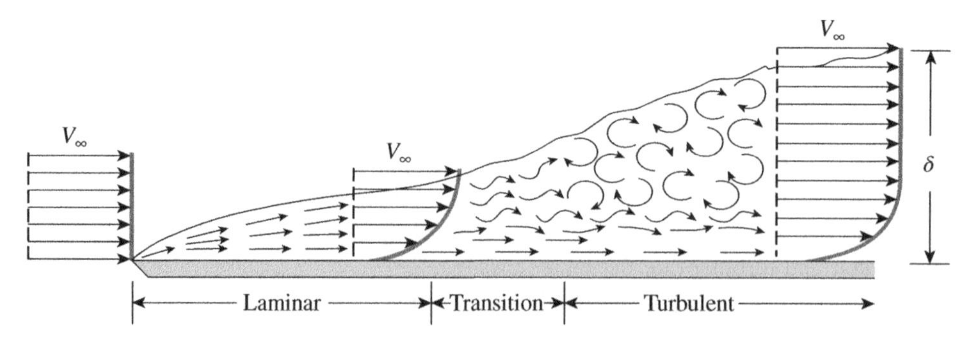


**Figure 7.1** Boundary layer development on a flat plate at zero angle of attack showing both laminar and turbulent regions. Note that at any $x$-location velocity profiles inside boundary layers begin at zero on the plate surface and rise smoothly to merge with the free stream at the free stream edge.

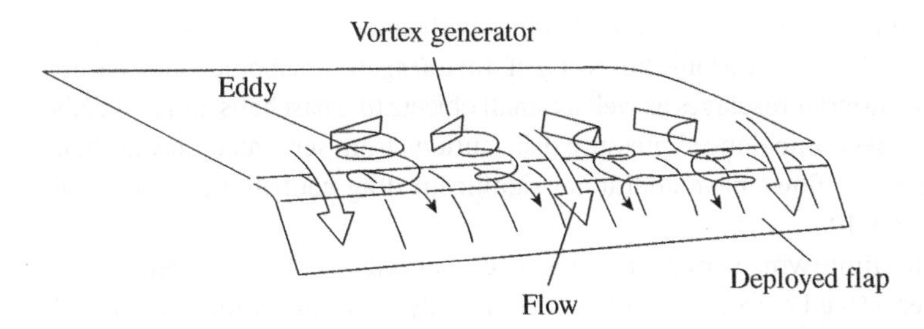


**Figure 7.2** Turbulence (vortex) generators mounted on a wing to delay separation during high-lift operation with deployed flaps. Vortex generators consist of tiny flow fences mounted on specific wing locations at several oblique orientations.

### 7.3.1 Viscous Flow Equations in Incompressible Flow

Because drag variations along the wing's span can be negligible, all flows considered in this chapter are two-dimensional with the $x$-direction along the free stream and $y$ (instead the $z$) as the cross-flow coordinate. The differential equations introduced in Chapter 2 are applied here with Euler's equation enhanced by a viscous component. For incompressible two-dimensional flow over a flat plate, we may write the equations of continuity and momentum as shown next where the latter only carries a viscous term that is active inside the boundary layer. Moreover, we may assume that all pressures reflect ideal flow conditions with pressures being impressed unchanged through the thin fluid layers (hence the total derivative). The boundary conditions $u$ and $v$ are zero when both $x$ and $y$ equal zero, and they remain zero at $y = 0$ (the no-slip condition) for all $x > 0$, whereas at $y = \delta$ (an arbitrarily defined boundary layer edge) $u = V_\infty$ and $v$ returns to zero. These conditions are implied in the velocity profiles depicted in Figure 7.1.

$$\frac{\partial u}{\partial x} + \frac{\partial v}{\partial y} = 0 \tag{7.1}$$

$$\rho u \frac{\partial u}{\partial x} + \rho v \frac{\partial v}{\partial y} = \mu \frac{\partial^2 u}{\partial y^2} - \frac{dp}{dx} \tag{7.2}$$

Equations (7.1) and (7.2) represent the well-known *Navier–Stokes equations* for laminar flow over a flat plate. They were solved by Blasius for $\alpha = 0°$, a flow condition where the pressure gradient term is zero. More general solutions that include the pressure gradient term over wedge shapes were done by Falkner and Skan (see Anderson 1978, 2017; Schlichting and Truckenbrodt 1979; Shevell 1983). To represent both drag and flow separation, in this chapter we will deal with approximate laminar-flow representations, first without the pressure term and then with it.

## 7.4 Contributions to Drag

Drag on flying objects arises from a variety of sources. Among the largest contributors to the drag are:

a) Skin friction (viscous shearing at all immersed surfaces)
b) Pressure distributions (form drag at the object's wake)
c) Induced or drag due to lift on finite wings generating lift
d) Wave drag (supersonic only) and other compressible flow drag sources

Most wings before stall along with other streamlined bodies with smooth contours show a predominance of drag due to *skin friction* in subsonic flows and of *wave drag* in supersonic flows. Large, thick bluff bodies in flight like aircraft fuselages as well as small objects like baseballs and golf balls exhibit mostly *pressure drag* due to unbalanced pressure distributions from separated flow at their wakes. Transonic and hypersonic flows have additional significant drag contributions, some of which are discussed in Chapter 10.

*Drag due to lift* results from a finite wing's shed trailing vortices that originate from the generation of lift as discussed in Chapter 6 (see Eq. (6.12)). While not commonly associated with viscosity, all vortices have a viscous origin. *Wave drag* is a strictly supersonic phenomenon that is treated in Chapters 9 and 10. Other contributions to the drag are individually less significant in well-designed systems and are deferred for more applied courses.

For incompressible flows it is convenient to divide the drag coefficient as defined in Chapter 2 into two parts, namely, zero-lift drag ($c_d$ or $C_{D0}$) and lift-induced drag ($C_{Di}$). Note that the sectional or two-dimensional drag coefficient $c_d$ is expected to be nearly the same as its finite or three-dimensional wing counterpart ($C_{D0}$) so that no distinction need be made, and we use the lowercase notation and uppercase drag notation interchangeably. As with other coefficient forms, we intentionally define drag coefficients to be dimensionless to enhance their generality.

$$C_D = c_d + C_{Di} \tag{7.3}$$

## 7.5 Skin-Friction Drag on Airfoils

Shear forces on an immersed object arise from the *shearing stress* $\tau$ applied by any flow parallel to its surface, in contrast to the pressure forces that act perpendicular to it (as depicted in Figures 2.4 and 4.2). The shearing stress, omitted from our ideal-flow momentum balance given in Eq. (2.10) and in Euler's equation, has been included in Eq. (4.13) as "losses" and in Eq. (7.2).

In direct analogy to the pressure coefficient, the shear or skin-friction coefficient is defined as [check the units!]:

$$C_f \equiv \frac{\tau}{\frac{1}{2}\rho_\infty V_\infty^2} \tag{7.4}$$

The total drag due to skin friction may be found by integrating the surface stress ($\tau$) first along the airfoil's chord length and then along its span, though the latter is unnecessary if there are no significant span variations. Using the *subscript u* for the upper and *l* for the lower surface, for a wing with planform $S$ we may write the drag per unit span:

$$c_d = \int_0^1 (C_{fu} + C_{fl})\, d\left(\frac{x}{c}\right) \tag{7.5}$$

$$c_d \approx C_{D0} = \frac{D_0}{q_\infty S} \tag{7.6}$$

In subsonic flows, skin friction on a lifting airfoil accounts somewhere between 30 and 50% of the total drag, whereas for non-lifting surfaces it accounts for all of the drag. In Figure 7.1 we have depicted a boundary layer on a smooth flat plate at $\alpha = 0°$ with its laminar and turbulent regions clearly evident. Because laminar flows produce less drag, airfoils are designed for cruising under mostly laminar conditions.

Satisfactory approximations exist for both laminar and turbulent flows over flat plates at $\alpha = 0°$, but none for the transition region that fortunately is relatively small. Because inside the boundary layers speed approaches the free stream asymptotically, a *boundary layer thickness* is defined arbitrarily at cross-flow location ($y$-distance) from the surface to where the local velocity has reached 99% $V_\infty$ (see Figure 7.1), and it is given the symbol ($\delta$). Equations (7.7) and (7.8) give empirical forms for both the boundary layer thickness and the drag coefficient ($C_f$) defined in Eq. (7.4) for a *flat plate at zero angle of attack* as a function of the local Reynolds number ($Re_x$). Note in these equations that both $\delta$ and $C_f$ depend on $x$. Another thickness similar to $\delta$, called the *displacement thickness*, is used to represent the overall movement or displacement of the ideal free stream by the presence of the slower flow region in the boundary layer and amounts to ($\delta/3$) in laminar flows and about ($\delta/8$) in turbulent flows for this case.

$$\text{Laminar: } \delta = \frac{5.2x}{\sqrt{Re_x}} \quad \text{and} \quad C_f = \frac{0.664}{\sqrt{Re_x}} \quad Re_x < 10^5 \tag{7.7}$$

$$\text{Turbulent: } \delta = \frac{0.37x}{(Re_x)^{0.2}} \quad \text{and} \quad C_f = \frac{0.0583}{(Re_x)^{0.2}} \quad Re_x > 10^5 \tag{7.8}$$

The Reynolds number as introduced in Chapter 2 was based on the airfoil's chord for overall conditions but here $Re_x$ is a variable based on the $x$-coordinate:

$$Re_x = \frac{\rho_\infty V_\infty x}{\mu_\infty} \tag{7.9}$$

Representative magnitudes of $Re$ encountered during flight are in the order of one million, so ($\delta/x$) is a very small number relative to other dimensions (e.g. with $Re = 10^6$ and $x = 30$ cm, $\delta =$ 0.16 cm for laminar flows and 0.70 cm for turbulent flows and the corresponding displacement thicknesses are even smaller). Figure 7.3 shows $C_f$ curves for flow over flat plates; notice the log-log scale in the figure. The transition Reynolds number shown is an average value related to flat plates. Because the transition region is generally short in length, we often apply simplifying assumptions for calculating sectional drag coefficients as shown in Example 7.1.

In order to find the *total skin-friction drag coefficients*, we need to integrate $C_f$ in Eqs. (7.7) and (7.8) over the entire flat-plate length "$c$" as shown in Eq. (7.5). Since we assume that any spanwise variation of the drag coefficient can be neglected, integration along "$b$" is unnecessary and the result in Eq. (7.6) applies to the entire airfoil. Note in the formulations shown next that the Reynolds number at $x = c$, being independent of $x$, has been factored out of the integrals so that all remaining variables are in non-dimensional powers of ($x/c$). Recall also that the total drag coefficient for a flat plate is $C_D \equiv D/q_\infty bc$ and that there is no drag due to lift for a flat plate at zero angle of attack – a condition also closely represented at small angles of attack.

Purely laminar flows over a flat plate:

$$c_d = C_{D0} = \left(\frac{0.664}{\sqrt{Re_c}}\right)\int_0^1 \left(\frac{x}{c}\right)^{-0.5} d\left(\frac{x}{c}\right) = \frac{1.328}{\sqrt{Re_c}} \tag{7.10}$$

Purely turbulent flows over a flat plate:

$$c_d = C_{D0} = \left(\frac{0.0583}{(Re_c)^{0.2}}\right)\int_0^1 \left(\frac{x}{c}\right)^{-0.2} d\left(\frac{x}{c}\right) = \frac{0.0729}{(Re_c)^{0.2}} \tag{7.11}$$

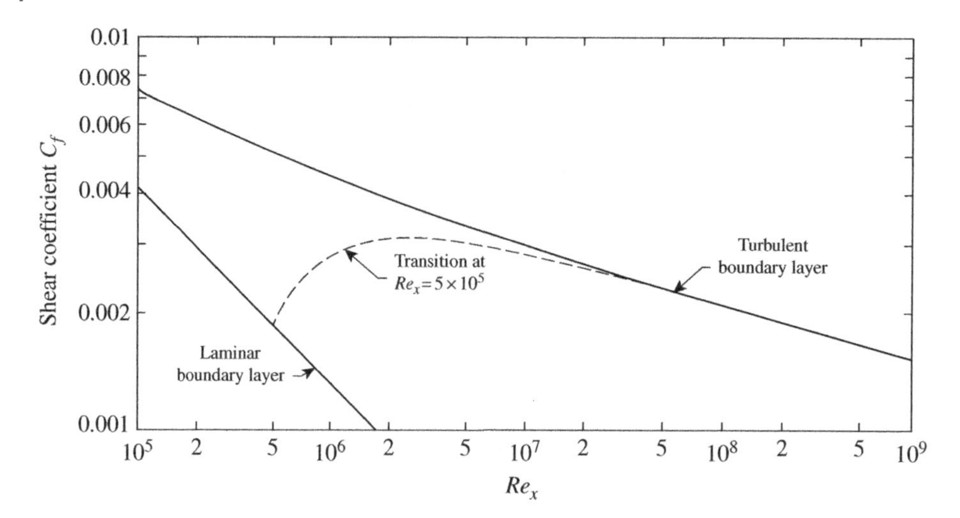


**Figure 7.3** Sectional drag coefficients on a flat plate at $\alpha = 0°$ as a function of $Re_x$. The laminar curve is from the Blasius solution and the turbulent curve is from work of Prandtl and von Karman. Transition is shown at only one of several typical values for the local Reynolds number.

**Example 7.1** A *flat plate* is 0.3 m long parallel to the free stream and 1.0 m wide perpendicular to it. Air at standard conditions flows at 60 m/s *over the upper side only.*

a) If the flow is entirely laminar, calculate the drag on the plate.
b) If the flow is entirely turbulent, calculate the drag.
c) If the transition Reynolds number is exactly at $5 \times 10^5$, what fraction of plate length would be under laminar flow?

At sea level: $\mu_\infty = 1.70 \times 10^{-5}$ kg/(m)(s) and $\rho_\infty = 1.226$ kg/m$^3$

$$Re_c = \frac{\rho_\infty V_\infty c}{\mu_\infty} = \frac{1.226 \cdot 60 \cdot 0.3}{1.70 \cdot 10^{-5}} = 1.30 \cdot 10^6 \text{ and } S = 0.30\,\text{m}^2$$

So the flow is partly laminar and partly turbulent. We may assume no flow variations along the span.

We can use Eqs. (7.10) and (7.11) to calculate the total drag on the plate.

a) Laminar only: $D = q_\infty S C_{D0} = 0.771$ N
b) Turbulent only: $D = q_\infty S C_{D0} = 2.11$ N
c) The laminar portion of the chord is about 0.12 m and about 0.18 m turbulent, assuming that the flowtransitions abruptly at $Re_x = 5 \times 10^5$. Since magnitude of the turbulent drag portion is greater, the answer in Part (b) should be closer to the overall drag.

For other than smooth flat plates it is best to consult the literature because much *empirical* information is available on sectional skin-friction coefficients. Measurements from a set of studies by the National Advisory Committee on Aeronautics are depicted in Figure 7.4 for the NACA 2412 airfoil section. Data show that for well-designed airfoils with attached boundary layers sectional drag coefficients are relatively constant only below $\alpha = \pm 8°$. During testing, these wing sections spanned an

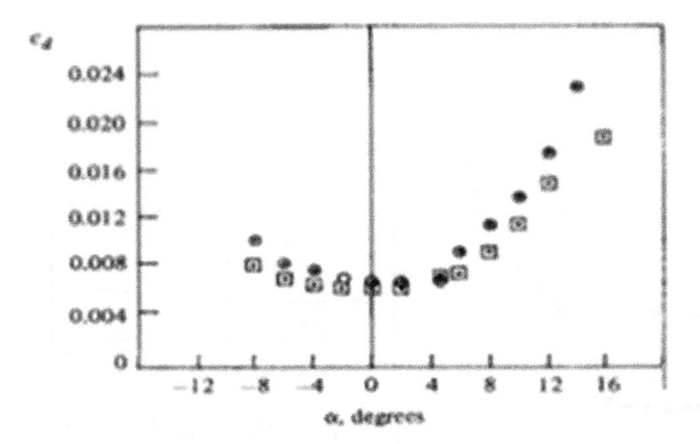


**Figure 7.4** NACA 2412 Sectional drag coefficient data at $Re_x = 3.1 \times 10^6$ (circles) and $8.9 \times 10^6$ (squares). *Source:* From Abbott and von Doenhoff (1949).

entire wind tunnel width, so there were no end effects due to finite widths (i.e. no drag due to lift), and these data can represent two-dimensional conditions.

**Example 7.2** Using the NACA 2412 airfoil sectional data shown in Figure 7.4, for $Re_x = 8.9 \times 10^6$ find $D_0$ at $\alpha = 0°$ and at the highest $C_L$ shown. The airfoil is rectangular with a chord length of 5 ft and a span of 20 ft. Assume that $c_d$ is constant along the span and that the flow operates in a standard atmosphere.

From data at this *Re*: $c_d = 0.006$ at $\alpha = 0°$ and $c_d = 0.018$ at 15°, which is the highest $C_L$ shown.

First we need to find the value of $V_\infty$ corresponding to the given Reynolds number using "standard conditions" for ($\rho_\infty$ and $\mu_\infty$):

$$V_\infty = \frac{Re_c \mu_\infty}{\rho_\infty c} = 280\,\text{ft/sec so that } q_\infty = \frac{\rho_\infty V_\infty^2}{2} = 93.2\,\text{slug ft}^2/\text{sec}^2 \quad (1\,\text{slug} = 32.2\,\text{lbm})$$

Since these data are two-dimensional, the $c_d$ is constant and the integrals in Eq. (7.6) become:

$$D_0 = c_d\, q_\infty\, S = 55.9\,\text{lbf at } 0° \quad \text{and} \quad 167.8\,\text{lbf at } 15°$$

Note that the total drag is about three times higher at the shown maximum value of the lift coefficient and that the Reynolds number effect on the curves is relatively small at the lower angles of attack (between − 8° and 8°). There is some error in assuming that the drag does not vary along the span, and when using the NACA data, most results are given as "drag per unit span."

Effects of airfoil thickness and surface roughness on the drag coefficient are shown with the data of Figure 7.5 for two other NACA airfoils. Of special interest is the appearance of a small "drag bucket" in the smooth airfoil just beyond $\alpha = 0°$ reflecting a limited decrease in drag coefficient. In this region completely laminar flow has been maintained and that is difficult to do in real aircraft because unavoidable surface contaminants (dust) and engine vibration trigger the transition to turbulence.

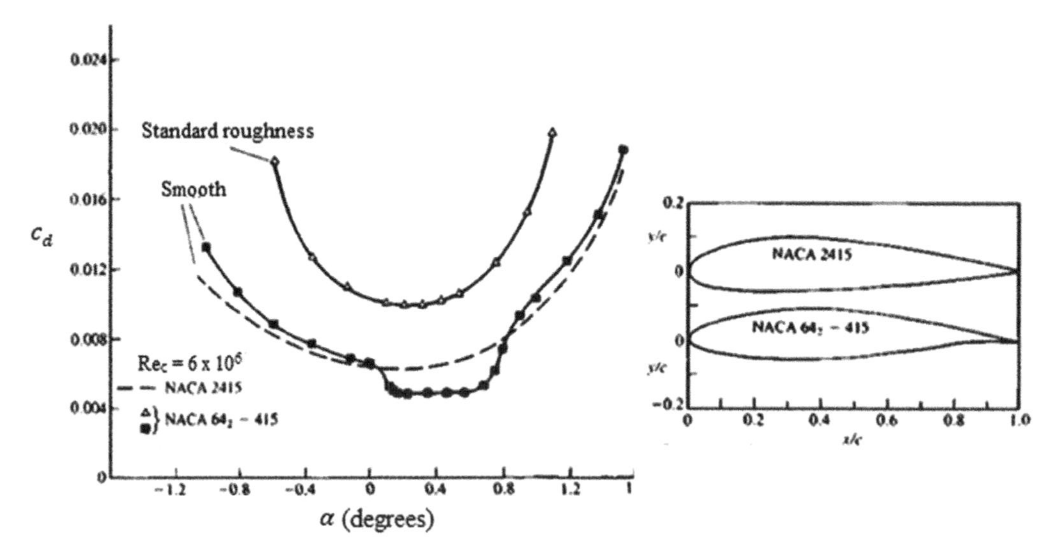


**Figure 7.5** Effects of thickness distribution and surface roughness on $c_d$ for the NACA 2415 (dashed line) and $64_2$-415 (solid lines, square = smooth and triangles = standard roughness) airfoils. The small *drag bucket* apparent in the lower curves reflects a purely laminar region that can only be maintained in wind tunnel tests. *Source:* From Abbott and von Doenhoff (1949).

## 7.6 Approximate Viscous Boundary Layer Profiles

Most of the viscous "action" in the boundary layer occurs next to the airfoil's surface where velocity gradients are highest, but where the velocity itself is very low enough so that the corresponding *local* Mach number remains below 0.3 (i.e. incompressible). This means skin-friction coefficient information shown in this chapter will apply even when free stream flow Mach numbers are in low supersonic ranges. Existing *analytical solutions* to the laminar boundary layer over a flat plate may be closely approximated with a velocity profile that consists of a "cubic parabola." While no turbulent-flow solutions have been purely analytical, the most common empirical equation uses a 1/7-power cross-flow coordinate profile. We proceed here with two approximate velocity-profile representations for a flat plate at zero angle of attack.

For the *laminar-flow* representation, we use the following approximate velocity profile:

$$\frac{u}{V_\infty} = \frac{3}{2}\left(\frac{y}{\delta}\right) - \frac{1}{2}\left(\frac{y}{\delta}\right)^3 \quad \text{where } 0 \le y \le \delta \tag{7.12}$$

which gives values within 3% of the exact (or Blasius) solution inside laminar boundary layers. The velocity profile $u$ is a function of both $x$ and $y$ in Eq. (7.12) since $\delta$ is a function of $x$ as in Eq. (7.7). Close to the wall equation (7.12) is quite accurate, and its principal defect shows up outside of the boundary layer (i.e. for $y > \delta$) where the $u$ vs. $y$ profile does not become the constant $V_\infty$ as it should.

For the *turbulent flow* representation, some time-averaged form of a time-dependent momentum equation is necessary because the flow is essentially *unsteady*. Here we will use an empirical profile called the *law of the wall* or a *1/7 power law* that is written as:

$$\frac{u}{u_m} = \left(\frac{y}{\delta}\right)^{1/7} \quad \text{where } 0 \le y \le \delta \tag{7.13}$$

In this equation, $u_m$ is the value of $u$ at $y = \delta$ that may be taken as $V_\infty$ for a flat plate at zero angle of attack. However, as seen in Figure 7.6, while the cubic parabola's slope goes to zero at $y = \delta$, the 1/7-velocity profile does not and keeps increasing very slightly at the expected edge of

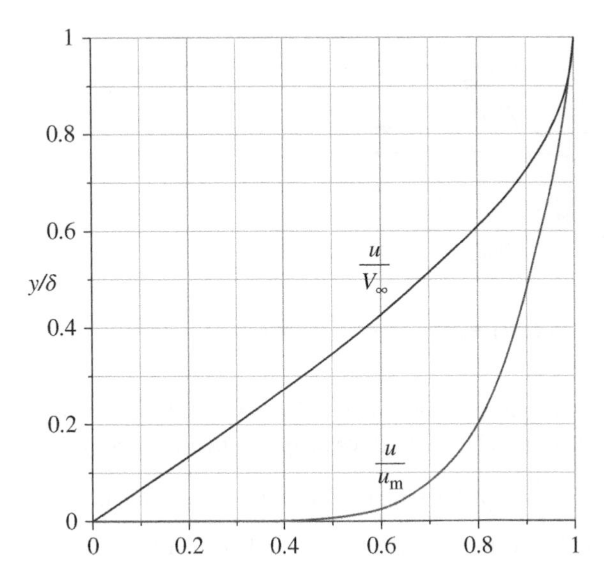


**Figure 7.6** Approximate velocity profiles in the cross-flow direction: laminar (top) and turbulent (bottom) as a function of $y/\delta$. Both profiles apply within the boundary layer region of a flat plate. Note the much steeper value of the slope $(\partial u/\partial y)$ for the turbulent profile next to the plate's surface that results in much higher turbulent drag coefficients.

the boundary layer (where both profiles are to be truncated). As already mentioned, the reason these two approximate profiles are useful is that they have proper curvatures deep in the boundary layers next to the wall where viscous effects are most significant.

### 7.6.1 Boundary Layer Separation

Conditions that lead to *flow separation* of laminar boundary layers are of special interest because these flows while producing less drag detach much more readily than turbulent ones. In order to examine flow separation in laminar boundary layers, we need to focus on the pressure gradient term on the right-hand side of Eq. (7.2) because it governs the behavior of the velocity. We are particularly interested in variables at the wall where $u = v = 0$ and Eq. (7.2) becomes:

$$\mu\left(\frac{\partial^2 u}{\partial y^2}\right)_{y=0} - \frac{dp}{dx} = 0 \tag{7.14}$$

Here $p$ is the free stream pressure that may only vary in the $x$-direction even throughout the boundary layer. The mathematical reason for surface detachment during flow separation is that:

$$\text{when } \frac{dp}{dx} > 0 \ \text{ then } \left(\frac{\partial^2 u}{\partial y^2}\right)_{y=0} > 0$$

and an *inflection point* or $\partial u/\partial y = 0$ eventually appears in the velocity profile at the wall in order to meet all necessary conditions. Analytical results (e.g. see Kuethe and Chow 1998; Schlichting and Truckenbrodt 1979) indicate that purely laminar boundary layers detach shortly after passing a sharp turn, such as over the deployed flap shown in Figure 5.10, at around $\eta \approx 15°$. Aft contours on airfoils have smooth curvatures as shown in Figure 7.7, and for such cases the boundary layer equations need to be solved numerically and/or airfoils need testing in a wind tunnel under the flow conditions experienced during flight.

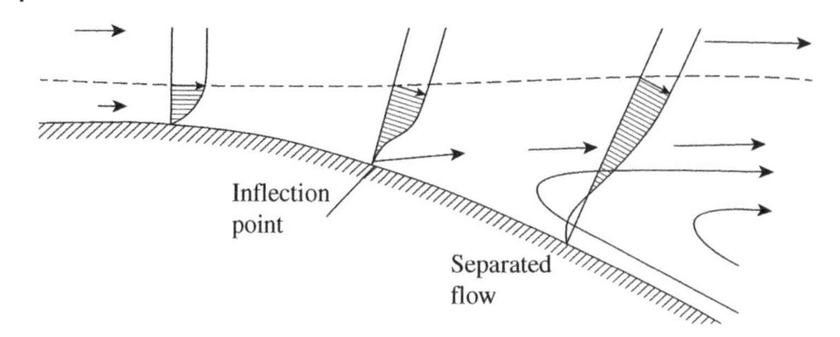


**Figure 7.7** Laminar boundary layer separation in an adverse pressure gradient region.

**Example 7.3** Let us demonstrate the mathematical existence of an *inflection point* by expanding the cubic parabola in our laminar-flow formulation to include values of $dp/dx$ other than zero. For this example we need to define a constant $k \equiv \frac{\delta^2}{\mu V_\infty}(dp/dx)$ to represent the pressure gradient term external to the boundary layer in Eq. (7.14). This new formulation will expand Eq. (7.12) by accommodating non-zero pressure gradients from Eq. (7.14) with an extra term, namely, $\frac{k}{2}\left(\frac{y}{\delta}\right)^2$. The expanded equation should also satisfy essential velocity profile boundary conditions becoming zero at $y/\delta = 0$ and one at $y/\delta = 1.0$. It is given as Eq. (E7.1).

$$\frac{u}{V_\infty} \approx \left[\frac{3}{2} - \frac{k}{4}\right]\left(\frac{y}{\delta}\right) + \frac{k}{2}\left(\frac{y}{\delta}\right)^2 - \frac{1}{2}\left[1 + \frac{k}{2}\right]\left(\frac{y}{\delta}\right)^3 \tag{E7.1}$$

As Figure E7.3 shows, at $k = +6.0$ the $u$-profile goes through an *inflection point* where $du/dy$ becomes zero at $y = 0$. This can also be shown by solving for $\frac{\partial u}{\partial y} = 0$ at the surface – it also turns

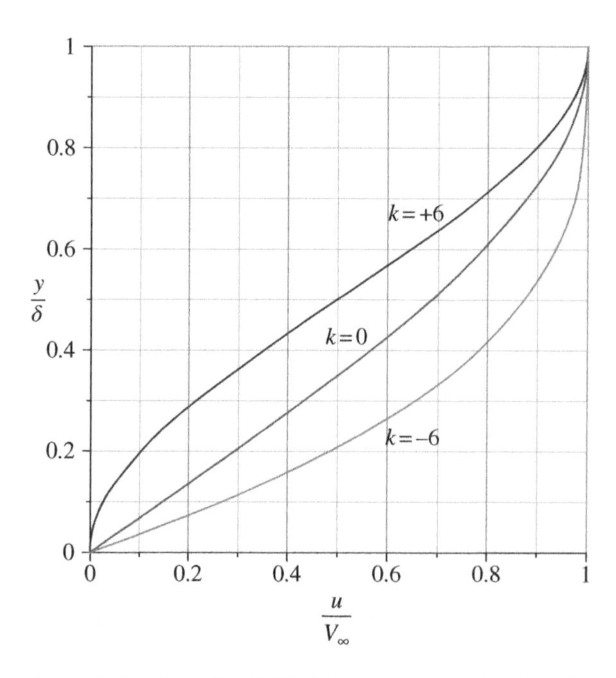


**Figure E7.3** Plot of Eq. (E7.1). Note, at the top inflection (+6) corresponding to a flow expansion, in the middle press gradient = 0, and at the bottom negative press gradient (−6) corresponding to a flow compression.

out that setting the multiplier of the first term in Eq. (E7.1) equals to zero also gives the value of the pressure gradient at the inflection point.

These results, though approximate, give correct pressure-gradient trends and Eq. (E7.1) has the attribute of defaulting to Eq. (7.12) for the velocity profile when there is no pressure gradient (the curve for $k = 0$). Although not built into the derivation, the slope of the $k = +6$ curve at $y/\delta = 1.0$ is relatively close to zero.

Results in this example compare favorably with Pohlhausen's *fourth-degree polynomial analysis* (see Kuethe and Chow 1998). The predicted inflection-point constant values differ by a factor of two, but note that both this and Pohlhausen's profiles are approximate.

## 7.7 Enrichment Topics

### 7.7.1 Air Brakes

Air brakes consist of devices built into wings to *increase drag*. They are only activated during flight maneuvers and especially during landing. They may be located at mid-portions or near the trailing edges of the wing; some can be operated through the same mechanisms as the flaps, see Figure 7.8. Air brakes work by locally inducing flow separation and are used to slow down as well as to control landing aircraft. Air brakes differ from *spoilers* in that air brakes are designed to increase drag while making little change on lift, whereas spoilers are used to noticeably reduce the lift-to-drag ratio. The end result (for the aircraft) is the same in that both help to slow down or intensify the rate of descent. Application of air brakes and spoilers eventually puts the full weight of the aircraft on the landing gear, so its conventional wheel brakes can be applied.

### 7.7.2 Hydrofoils

As they move through water, sailboats experience much more resistance than sail ice-boats while being propelled by wind forces as described in Section 6.9. A hydrofoil is a hydrodynamic lifting device, or foil, that operates underwater generating lift to raise watercraft out of water – unlike

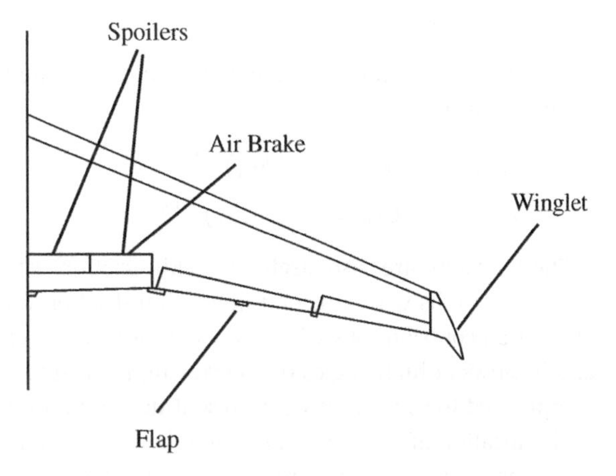


**Figure 7.8** Typical locations for air brakes and spoilers on wing.

aircraft, boats being neutrally buoyant do not need lift forces to stay afloat. Hydrofoils are used in sailboats for racing and on highspeed watercraft for reducing overall drag. Since traveling in air and water are governed by the same incompressible flow principles (but with different amounts of viscosity and density), the hydrofoil generates lift in the same manner an airfoil does while operating fully immersed in water. When the weight of the moving boat is fully balanced by the lift being generated, the hydrofoil vessel has reached *cruising* mode with its main structure above the water. While in air, there is no wave resistance on the hull and skin drag on the hull is appreciably lessened, with higher resistance water flow only affecting the much smaller contact areas of the hydrofoil. This results in a noticeable speed advantage with vessels being propelled by wind power (see Problem 7.4) and lessens engine needs in highspeed boats. Stability and control of watercraft with hydrofoils can become complicated, and this makes racing boats such an interesting sport.

## 7.8 Summary

Drag from skin friction and from shed wakes originates from the viscous nature of fluids. Vorticity is always present inside boundary layers causing some dissipation of flow energy into heat (which becomes most noticeable in hypersonic flows, i.e. when $Ma_\infty > 5$). In low viscosity fluids such as atmospheric air, the flow most affected by shear forces remains very close to the moving object, allowing all regions outside of the boundary layers to satisfy ideal flow assumptions while not truly inviscid. Pressure drag also originates in the boundary layers from flow that encounters sufficiently adverse pressure gradients to detach, resulting in a separated region or wake. Stall in well-streamlined objects such as airfoils is most apparent as angles of attack above 15°.

In this chapter we discussed several effects of viscosity that affect airfoil operation. We focused on some particulars of the flow inside attached boundary layers. For both "ribbon-like" or laminar flows and "highly mixed" or turbulent flows, we identified approximate velocity profiles and assumed that the transition region between these two regimes to be relatively short. The local boundary layer height or thickness ($\delta$) and the local shear or skin-friction coefficient (defined as $C_f = \tau/\frac{1}{2}\rho V_\infty^2$) for smooth flat plates at zero angle of attack in the laminar and turbulent regions are given by:

$$\text{Laminar: } \delta = 5.2x(Re_x)^{-0.5} \quad \text{and} \quad C_f = 0.664(Re_x)^{-0.5} \quad Re_x < 10^5 \tag{7.7}$$

$$\text{Turbulent: } \delta = 0.37x(Re_x)^{-0.2} \quad \text{and} \quad C_f = 0.0583(Re_x)^{-0.2} \quad Re_x > 10^5 \tag{7.8}$$

In either fully laminar or fully turbulent flows, the *total* skin-friction drag coefficients on a flat plate are given by:

$$\text{Laminar: } C_{D0} = 1.328(Re_c)^{-0.5} \tag{7.10}$$

$$\text{Turbulent: } C_{D0} = 0.0729(Re_c)^{-0.2} \tag{7.11}$$

These results are also useful for skin-friction estimates in incompressible flows over well-designed airfoils with attached flows. Skin-friction drag will be shown to remain comparable to other drag contributions when flow compressibility effects become important, in transonic flows and in flows at high angles of attack. Empirical studies are necessary here as theory remains too complicated to solve even with present-day computers.

The location of the transition region between laminar and turbulent can be controlled through increased surface mixing introduced with "turbulence generators" – turbulent flows possess energetic cross-flow mixing, which is why they detach less easily, but they increase skin friction so their extent is minimized as much as possible. Air brakes are designed to create drag while spoilers are designed to destroy lift. Flow separation criteria and various other particulars of viscous-flow-like transition to turbulence, while important, are beyond the scope of our presentation.

## Problems

Note: Assume that transition to turbulence occurs abruptly at $R_{ex} = 5 \times 10^5$ for flat plates.

**7.1** Calculate the total drag coefficient, $C_{D0}$, on a flat-plate airfoil moving at 86 m/s at a height of 2.4 km. The chord length is 2 m and the total span length is 3 m. Assume boundary layers with negligible spanwise flow and either:
i) all laminar.
ii) all turbulent.

**7.2** Consider a two-dimensional flat plate at zero angle of attack in a standard sea-level atmosphere with uniform flow at $V_\infty = 20$ m/s. If the total drag coefficient has equal contributions from the laminar and turbulent portions of the boundary layer, what is the length of the plate?

**7.3** Two closely spaced parallel flat plates are immersed in still air. The upper plate is moving with respect to the lower plate with a velocity $U$. Give the velocity profile that develops in terms of $y$, where $0 \leq y \leq \delta$. What is the relationship for the shearing stress (force per unit area) at the lower wall for this flow if the viscosity is $\mu$? (Hint: The no-slip condition forces the fluid to move at the same speed as the plate at each contact surface and the flow is laminar.)

**7.4** What modifications are necessary to the lift and drag equations developed for air flows to apply to hydrofoils that operate in water? Would the existing dimensionless lift and drag coefficient data on airfoils be useful, and what would be the role of the Reynolds number? (Hint: At ambient conditions the viscosity of water is about 40 times greater than air, whereas the kinematic viscosity that governs the Reynolds number as shown in Figure 7.3 is about 30 times less.)

**7.5** Incompressible flow measurements of lift and drag are to be conducted on a prototype airfoil, and the question is whether to use a wind tunnel or a water tunnel. Assume that the air available is at 15 °C and the water at 20 °C. We need to match $C_L$ and $Re$ in both test setups using the same airfoil. Calculate the speed ratio as well as the ratio of lift forces that would be measured between the air and the water tunnel tests.

**7.6** Consider a wing with an elliptic planform flying at sea-level standard conditions. The free stream velocity is 30 m/s. In order to calculate the drag due to skin friction, consider an equivalent rectangular plate planform of chord length 1.5 m and aspect ratio of 5 at $\alpha = 0°$. Calculate the total drag coefficient for this airfoil when $C_L = 0.8$ (assuming the airfoil has not stalled).

**7.7** Skin friction together with induced drag account for most of an aircraft's drag. One method of decreasing the induced drag is to increase the aspect ratio (***AR***) of the wing. Showing appropriate arguments, discuss the effect of increasing the ***AR*** on the total drag for the cases given next. You may assume a rectangular flat plate at $\alpha = 0°$ for skin-friction calculations and same weight and speed at the desired cruising altitude.

a) Increase the ***AR*** but keep the planform area constant.
b) Change the ***AR*** by increasing the chord keeping the span constant.
c) What limitations other than aerodynamic might restrict such "slendering" of a wing?

**7.8** The walls of a wind tunnel can be made slightly divergent to offset the effects of boundary layer growth that reduces its constant speed cross-section along the test length. At what angle should the wind tunnel walls be set so that the boundary layer does not encroach into a testing region. Use an effective Reynolds number of $10^6$.

## Check Test

**7.1** Describe in your own words the difference between laminar and turbulent flow and the desirability of each type for airfoil operation.

**7.2** In what types of flows may Bernoulli's equation *not be used* to calculate the pressure over wing?

**7.3** Any physically correct velocity profile across a boundary layer should be zero at the surface and approach the free stream with zero first and second derivatives. Do the approximate profiles in Eqs. (7.12) and (7.13) meet *both* of these requirements?

**7.4** Two identical wings are flying in tandem (i.e. wing B flying directly behind wing A). The wings are flying at the same speed. Discuss considering them separately the:

a) possible effects of velocities induced by one wing on the other.
b) effect of the wake of wing A on wing B; also discuss which wing requires greater propulsive power to fly at the same speed.

**7.5** State the dimensionless parameters that characterize viscous flow and give values for typical air flows.

**7.6** What speed boundary conditions must apply at the surface of an immersed object for:

a) viscous flow?
b) inviscid flow?

# 8

# Fundamentals of Compressible Flow

## 8.1 Introduction

In the last three chapters, we concentrated on incompressible flows since they are the most common in aerodynamics. We found that when we can treat air as constant density fluid, there are many similarities between aerodynamics and hydrodynamics – the governing equations are essentially the same and they have common dimensionless coefficients. Differences between flows in air and water only become apparent when we apply values for density and viscosity in the equations. Compressibility begins to affect flows as the Mach number (*Ma*) increases past 0.3 and turns out to be most obvious at supersonic speeds where we will focus most of our work after we develop the governing equations in ideal compressible flows. While subsonic flows can detect objects before encountering them, supersonic flows cannot and in order to turn they create a pair of unique flow conditions. In this chapter we discuss how purely supersonic phenomena such as shocks (supersonic compressions) and Prandtl–Meyer waves (supersonic expansions) enable flow turning and affect the nature of lift and drag on airfoils. We will then analyze a slender flat plate under supersonic flight since it is the most desirable configuration in this flow regime.

## 8.2 Objectives

After successfully completing this chapter, you should be able to:

1) Define the speed of sound and state the equation associated with it for perfect gases.
2) Define the stagnation state and its usefulness in isentropic flows.
3) Write the equation for the stagnation temperature ($T_t$) of a perfect gas in terms of the static temperature ($T$), Mach number ($Ma$), and ratio of specific heats ($\gamma$).
4) Write the equation for the stagnation pressure ($p_t$) of a perfect gas in terms of the static pressure ($p$), Mach number ($Ma$), and ratio of specific heats ($\gamma$).
5) (*Optional*) Show the steps involved in deriving Eqs. (8.7) and (8.9) from fundamentals as given in Chapters 1 and 4.
6) Describe in your own words how turning over a convex corner in supersonic flow can be isentropic.
7) Define the Prandtl–Meyer *expansion* and its associated flow effects.
8) State in words what Eq. (8.12) signifies and why $Ma = 1.0$ is relevant as a reference point.
9) Describe the *normal shock* and state if there are any associated flow turning conditions.

*Elements of Aerodynamics: A Concise Introduction to Physical Concepts*, First Edition. Oscar Biblarz.

Companion website: www.wiley.com/go/elementsofaerodynamics

10) Explain how an *oblique shock* results from the superposition of a normal shock with flow displacement tangential to the shock.
11) Sketch an oblique shock and define *shock angle* and *deflection angle.*
12) State in words what Eq. (8.20) signifies and why external shocks need to be weak shocks.
13) Calculate the pressure forces on a flat plate at angle of attack in supersonic flows.

## 8.3 Speed of Sound and Mach Waves

We begin by examining how pressure disturbances propagate through an *elastic medium* such as air. At any given point in a gas, disturbances create a region of compressed molecules that is passed along to neighboring regions in the form of a *traveling wave.* Waves come in various strengths depending on the amplitude of the pressure disturbance. The speed at which pressure disturbances propagate through the medium is called the *wave speed.* This speed is related to both the type of medium and its thermodynamic state, in addition to being a function of the strength of the wave. The *stronger* the wave, the *faster* it moves, and when strong enough these waves can affect the gas temperature of the medium they travel through.

The speed of sound ($a$) is the speed under which *infinitesimally weak pressure waves* travel through a substance be it a gas, liquid, or solid. It is a property of the medium, and for air treated as a perfect gas, $a^2$ may be written as (see Section 1.4):

$$a^2 = (dp/d\rho)_s = \gamma RT \tag{8.1}$$

In Eq. (8.1) $\gamma$ is the ratio of specific heats and $R$ the gas constant, both introduced in Chapter 1. The pressure derivative with respect to density is shown here as isentropic because very weak pressure waves do not affect the properties of the medium through which they propagate. Stronger pressure waves, or shock waves, arise from objects traveling faster than the speed of sound (i.e. supersonically). Figure 8.1a depicts the propagation of a sound wave, and in Figure 8.1b we represent the triangle that embodies the *Mach angle* ($\mu$) definition where:

$$\sin(\mu) = 1/Ma = a/V$$

Here $Ma$ is the Mach number ($Ma = V/\sqrt{\gamma RT}$). The radius of each circle in Figure 8.1a increases according to the product of some fixed time increment and the wave speed. Note that $\mu$ does not exist in subsonic flows and that it has a value of 90° at Mach 1.0, decreasing in magnitude as the Mach number increases. The two slanted solid lines (one labeled "Mach cone") emanating from point P are called "characteristics" since they are related to the local Mach number. Flow streamlines cross these characteristic lines unchanged because they do not affect gas properties. This is in contrast to flow crossing a shock that will be shown to compress and potentially turn the flow depending on surface orientation and boundary conditions. The direction of any characteristic line, also called a Mach line, is found from the triangle in Figure 8.1b from

$$\cot^2\mu = Ma^2 - 1.0$$

Mach lines actually issuing from a point define a *conical surface* with vertex at "P," but in Figure 8.1a we have only drawn their projection on a plane.

Two-dimensional representations of supersonic flows approaching from the left are always drawn with two downstream-moving families of Mach lines – directed down and up. They are named characteristics because of the hyperbolic nature of the differential equations that describe

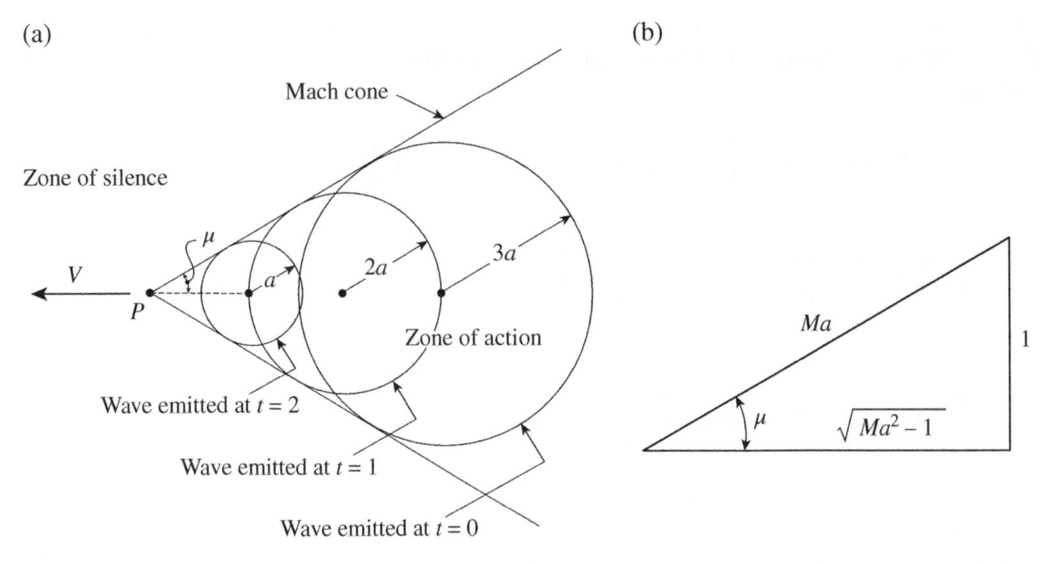


**Figure 8.1** (a) Wave fronts from a weak supersonic disturbance; (b) trigonometric relation between Mach number (*Ma*) and Mach angle ($\mu$).

supersonic flows. These characteristic lines have a well-defined direction (which corresponds to increasing time in the "wave equation"). Physically, this means that supersonic flows cannot detect upstream conditions in contrast to subsonic ones.

**Example 8.1** Calculate the temperature and density for air at a location where the speed of sound is 995.8 ft/sec and the pressure is 628.45 lbf/ft$^2$.

We will assume that air behaves as a perfect gas. To use the information given, we need to account for fact that EE units require the unit-conversion factor $g_c = 32.174$ lbm-ft/lbf-sec$^2$. For air $\gamma = 1.4$ and $R = 53.3$ ft-lbf/lbm-°R.

$$T = \frac{a^2}{\gamma g_c R} = \frac{998.5^2}{1.4 \times 32.174 \times 53.3} = 415.25^\circ\text{R} = -44.39^\circ\text{F}$$

$$\rho = \frac{p}{RT} = \frac{628.45}{53.3 \times 415.25} = 0.0284\,\text{lbm/ft}^3$$

## 8.4 Steady-State Isentropic Flow

We are now ready to examine details of compressible flows approaching a wing. This type of movement is treated as ideal and, in the language of thermodynamics, isentropic; the applicable equations were introduced in Chapter 1 for perfect gases with constant $\gamma$. We now proceed to develop formulas that describe compressible flows so as to replace Bernoulli's equation.

In order to arrive at working equations for perfect gases in compressible flows, we will use the speed of sound for perfect gases, Eq. (8.1), the perfect gas equation of state ($p = \rho RT$), and the definition of the Mach number. We will want to integrate Euler's momentum formulation, Eq. (4.2) with a variable density between a static and a stagnation state (applying Eq. (1.7) with $n = \gamma$) in Eq. (8.2). Equations (8.3)–(8.5) show intermediate steps required to arrive at Eq. (8.5), which shows the resulting speed of sound between the aforementioned static and stagnation states for irrotational but compressible gas flows. Equation (8.6) is conveniently written in terms of absolute gas temperatures and the Mach number.

$$\text{Equation (4.2)}\quad \frac{V^2}{2} + \int_p^{p_t} \frac{dp}{\rho} \quad \text{is integrated with } \rho = \rho_t \left(\frac{p}{p_t}\right)^{\frac{1}{\gamma}} \tag{8.2}$$

$$\int_p^{p_t} \frac{dp}{\rho} = \frac{p_t^{\frac{1}{\gamma}}}{\rho_t} \int_p^{p_t} \frac{dp}{p^{\gamma}} = \frac{\gamma \frac{p_t^{\frac{1}{\gamma}}}{\rho_t}}{\gamma - 1}\left[p_t^{1-\frac{1}{\gamma}} - p^{1-\frac{1}{\gamma}}\right] = \frac{\gamma R(T_t - T)}{\gamma - 1} \tag{8.3}$$

but since

$$\left(\frac{p_t^{\frac{1}{\gamma}}}{\rho_t}\right)\left(p_t^{1-\frac{1}{\gamma}}\right) = RT_t \quad \text{and} \quad \left(\frac{p_t^{\frac{1}{\gamma}}}{\rho_t}\right)\left(p^{1-\frac{1}{\gamma}}\right) = RT \tag{8.4}$$

substituting in Eqs. (8.2) and (8.1)

$$\frac{V^2}{2} + \frac{a^2}{\gamma - 1} = \frac{a_t^2}{\gamma - 1} \quad \text{dividing next by } \frac{a^2}{\gamma - 1} \tag{8.5}$$

$$\frac{\gamma - 1}{2} Ma^2 + 1 = \frac{a_t^2}{a^2} = \frac{T_t}{T} \tag{8.6}$$

Solving Eq. (8.6) for $(T/T_t)$ leads to its more common form as Eq. (8.7) shown next. The variables with subscript "$t$" denote a stagnation condition where the velocity is zero, and for our gas flows *total* and *stagnation* conditions are equivalent since potential energy has been neglected. We can now write the companion isentropic perfect gas equations for the density and pressure ratios as Eqs. (8.8) and (8.9). With the Mach number ($Ma$) and the ratio of specific heats ($\gamma$) shown explicitly, we have arrived at a well-known form of the *isentropic perfect gas relations*. These ratios are depicted in Figure 8.2.

$$\frac{T}{T_t} = \left(1 + \frac{\gamma - 1}{2} Ma^2\right)^{-1} \tag{8.7}$$

$$\frac{\rho}{\rho_t} = \left(\frac{T}{T_t}\right)^{\left(\frac{1}{\gamma - 1}\right)} = \left(1 + \frac{\gamma - 1}{2} Ma^2\right)^{-\left(\frac{1}{\gamma - 1}\right)} \tag{8.8}$$

$$\frac{p}{p_t} = \left(\frac{T}{T_t}\right)^{\left(\frac{\gamma}{\gamma - 1}\right)} = \left(1 + \frac{\gamma - 1}{2} Ma^2\right)^{-\left(\frac{\gamma}{\gamma - 1}\right)} \tag{8.9}$$

Using Eq. (8.9), the *stagnation-condition pressure coefficient* for the free stream in compressible flows may also be written in terms of $Ma_\infty$ and $\gamma$ as:

$$C_{pt} = \frac{p_t - p_\infty}{\frac{1}{2}\rho_\infty V_\infty^2} = \frac{2}{\gamma Ma_\infty^2}\left[\left(1 + \frac{\gamma - 1}{2} Ma_\infty^2\right)^{\left(\frac{\gamma}{\gamma - 1}\right)} - 1\right] \tag{8.10}$$

## 8.5 Supersonic Flows

When an object moves *subsonically* surrounding fluid particles can sense its position because it always moves within the "zone of action" (the circle of emitted waves at given times) shown in Figure 8.1, but when an object moves *supersonically* the approaching flow cannot sense anything

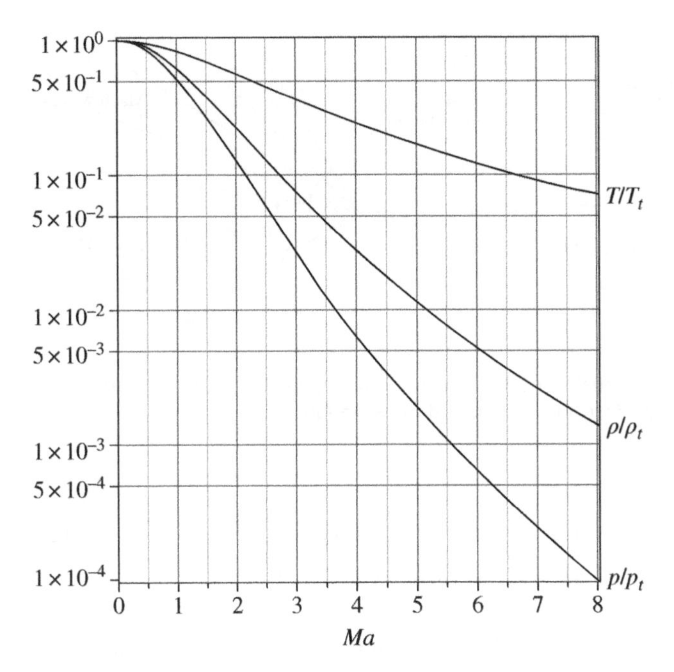


**Figure 8.2** *Isentropic ratios* of $T/T_t$, $\rho/\rho_t$, and $p/p_t$ vs. Mach number *Ma* for $\gamma$ = 1.4. The subscript "*t*" indicates the stagnation condition.

ahead of the moving object. This is called the "zone of silence," also shown in Figure 8.1. Supersonic flow streamlines therefore only turn when they encounter a barrier in their path, and there are two different ways supersonic flows can turn – one is through an expansion (convex turn) and the other through an oblique shock (concave turn). Expansions or convex turning occur mostly in external flows and generate no losses. Shocks are narrow compression zones, and common examples are normal shocks inside supersonic nozzles that do not turn the flow but do result in substantial losses. Supersonic expansions, also called Prandtl–Meyer waves, are covered first.

### 8.5.1 Prandtl–Meyer Flows (Isentropic)

It can be shown that convex turning (supersonic expansion) is an isentropic process and that such expansions occur along the Mach waves or "characteristics" as shown in Figure 8.3. The centered expansion on Figure 8.3b is thermodynamically equivalent to the turning in Figure 8.3a and more conveniently analyzed as here the flow is fully parallel to each wall before and after the expansion fan where it may be treated one dimensionally.

For the centered expansion geometry of Figure 8.3b, after assigning a new symbol ($\nu$) to the differential change in the Mach angle resulting from the expansion and some manipulation, we can write:

$$d\nu = \sqrt{Ma^2 - 1}\frac{dV}{V} \tag{8.11}$$

Here $d\nu$ represents the angular increment from an oncoming velocity vector that corresponds to the direction of $V$. After Eq. (8.11) is integrated and after setting $\nu = 0$ when $Ma = 1.0$, we arrive at the *Prandtl–Meyer function*:

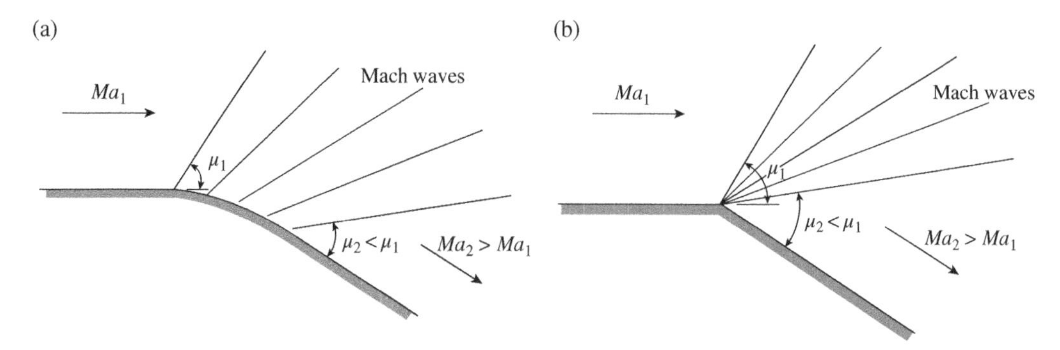


**Figure 8.3** Expansion waves: (a) two-dimensional smooth corner turn and (b) locally one-dimensional sharp corner "two segment equivalents" flow (except within the Mach wave fan).

$$\boxed{\nu = \left(\frac{\gamma+1}{\gamma-1}\right)^{1/2} \tan^{-1}\left[\frac{\gamma+1}{\gamma-1}\left(Ma^2-1\right)\right]^{1/2} - \tan^{-1}\left(Ma^2-1\right)^{1/2}} \tag{8.12}$$

Note in Eq. (8.12) that for any given $\gamma$ and supersonic Mach number, we have a unique function, $\nu = f(\gamma, Ma)$, one that is often tabulated. In order to find the Mach number after an expansion ($Ma_2$), we solve for $\nu_2$ by adding the $\Delta\nu$-change from the physical turn to the initial Prandtl–Meyer angle ($\nu_1$) as if the entire change occurred at the centered expansion, namely,

$$\nu_2 = \nu_1 + \Delta\nu \tag{8.13}$$

Knowing $Ma_1$ and $Ma_2$ we apply the isentropic relation, Eq. 8.9, to obtain the pressure ratios across the expansion; it is important to reiterate that stagnation pressure does not change since expansion turns are isentropic. Figure 8.4 may be used to estimate the pressure ratio $p/p_t$ at each Mach number or $\nu$-value. *Prandtl–Meyer* results are directly obtainable and with better *Aerodynamics Calculator* for air and other ideal gases in a larger range of parameters and with better accuracy.

Prandtl–Meyer expansion pressures as given in isentropic tabulations are ratioed to the total or stagnation pressure as per equation (8.9). To calculate the pressure coefficient between an incoming flow (station (1) in Figure 8.3) and outgoing flow (station (2) in Figure 8.3), we can write the definition given in Chapter 2 to read as follows:

$$C_p = \frac{\left(\frac{p_2}{p_t}\right) - \left(\frac{p_1}{p_t}\right)}{\left(\frac{\gamma}{2}\right)\left(\frac{p_1}{p_t}\right) Ma_1^2}$$

## 8.5.2 Oblique Shocks (Non-isentropic)

Shock waves are very narrow pressure jumps that develop in supersonic flows. Inside supersonic nozzles they arise from the need to match an exit-pressure increase. In external flows, however, they are triggered by compressive turns around an object. In aerodynamic flows, we are mostly concerned with *oblique shocks*; these are oriented somewhere between a Mach wave (for shocks of zero

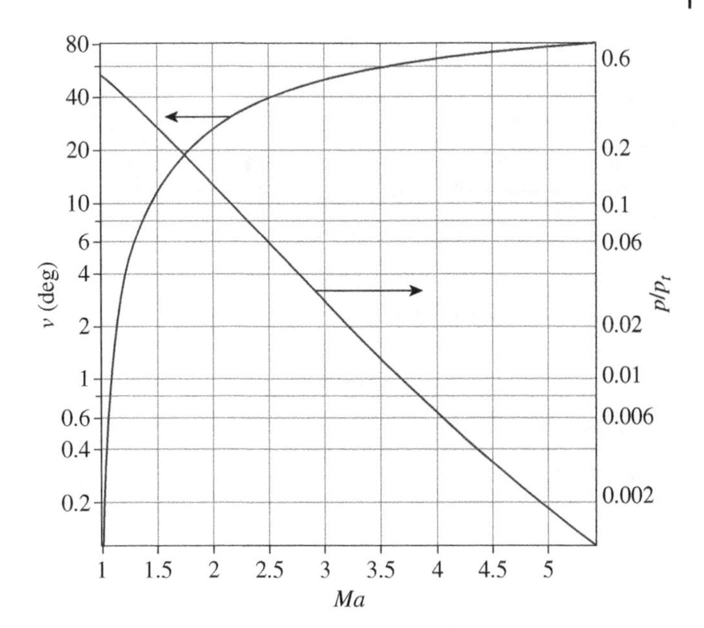


**Figure 8.4** Prandtl–Meyer expansion angle $\nu$ (deg) and $p/p_t$ versus $Ma$ for $\gamma = 1.4$.

strength) and the direction normal to the flow velocity (for shocks of maximum strength) – neither of these extremes turns the flow but in between at each turning angle ($\delta$) there are two possible values of shock angle ($\theta$). Oblique shocks are non-isentropic and require the simultaneous solution of the gas dynamic equations given in Chapter 2, namely, continuity, momentum, and energy. These equations are specialized for no-area changes since flow streamlines only travel through a narrow shock region. The flow is treated as ideal before and after the shock, but some flow-energy dissipation does occur within the shock.

As shown in Figure 8.5, the approaching free stream velocity component normal to the oblique shock becomes $V_{1n} = V_1 \sin\theta$ where $\mu_1 \le \theta \le 90°$ is the oblique shock angle.

Since there are no area changes, in the following equations cross-sectional areas across the shock have been cancelled out (i.e. $A_1 = A_2$).

$$\rho_1 V_{1n} = \rho_2 V_{2n} \qquad \text{continuity} \tag{8.14}$$

$$p_1 + \rho_1 V_{1n}^2 = p_2 + \rho_2 V_{2n}^2 \qquad \text{momentum (normal)} \tag{8.15}$$

$$V_{1t} = V_{2t} \equiv V_t \qquad \text{tangential velocities} \tag{8.16}$$

$$h_1 + 1/2\left(V_{1n}^2 + V_t^2\right) = h_2 + 1/2\left(V_{2n}^2 + V_t^2\right) \qquad \text{energy} \tag{8.17}$$

All property changes across oblique shocks are the same as those across an equivalent normal shock at $V_{1n}$, which shares the same equations but with $\theta = 90°$. The tangential velocity present creates the "inclination" of the total velocity vector. Referring back to Figure 8.5, a flow-deflection turn results from the smaller magnitude of $V_{2n}$ with respect to $V_{1n}$ since $V_t$ remains unchanged. The angle $\delta$ rotates the flow toward the shock so as to make $V_2$ parallel with the direction of the new boundary.

After some manipulation of the governing equations, we arrive at Eqs. (8.18) and (8.19) for the Mach numbers before and after the shock. It can be shown that the governing relation between

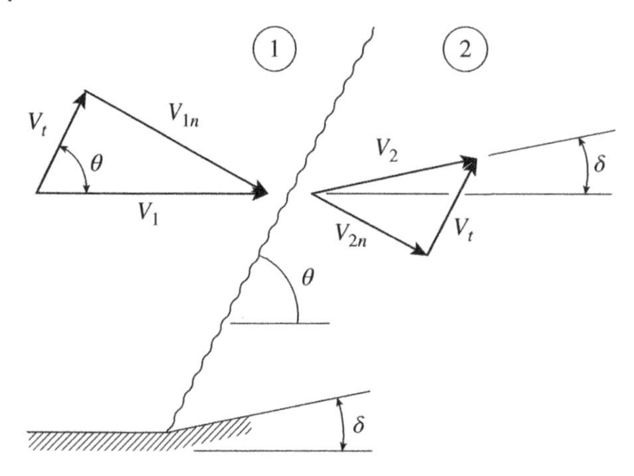


**Figure 8.5** Oblique shock with relevant angle definitions. The oblique shock angle is $\theta$ and the flow needs to turn inwards by an angle $\delta$ to remain parallel to the wall after the shock.

$\theta$ and $\delta$ is given by Eq. (8.20) where $\gamma$ is the ratio of specific heats and $Ma_1$ the approaching Mach number. Graphs of $\delta$ from Eq. (8.20) indicate that it goes through a maximum when plotted with respect to $\theta$ at any given supersonic Mach number (see Anderson 2017; Bertin and Cummings 2013; Kuethe and Chow 1998; McCormick 1979). As stated earlier, there can be two values of $\theta$ for each $\delta$ – one corresponding to a weak oblique shock and the other to a strong one. In external flows only *attached weak shocks* are known to take place. Detached normal shocks are produced when the flow encounters a blunt object head-on, and such shocks eventually turn into oblique shocks at the object's sides.

$$Ma_{1n} = Ma_1 \sin\theta \tag{8.18}$$

$$Ma_2 = \frac{Ma_{2n}}{\sin(\theta - \delta)} \tag{8.19}$$

$$\boxed{\tan\delta = 2(\cot\theta)\left(\frac{Ma_1^2 \sin^2\theta - 1}{Ma_1^2(\gamma + \cos 2\theta) + 2}\right)} \tag{8.20}$$

For thin and symmetric supersonic airfoils, the turning angle $\delta$ is very closely equal to the angle of attack $\alpha$ (i.e. $\delta \approx \alpha$), so to find the angle $\theta$ that corresponds to a weak oblique shock the only other information needed is $Ma_1$ and $\gamma$. Knowing $\theta$, property ratios across the shock can then be solved for. Equations (8.21) and (8.22) give pressure and temperature ratios across the shock, where condition (1) is before the shock and condition (2) after. Figure 8.6 shows resulting static pressure ratios ($p_2/p_1$) across oblique shocks as a function of turning angle $\delta$ for various approaching Mach numbers – the nearly horizontal portions of these curves correspond to the strong-shock portion of Eq. (8.20). The dual value of $p_2/p_1$ in Figure 8.6 at every turning angle $\delta$ derives from the dual value of $\theta$ from Eq. (8.20) already mentioned. In weak shocks with $\delta < 3°$ the pressure ratios are seen to be fairly independent of Mach number.

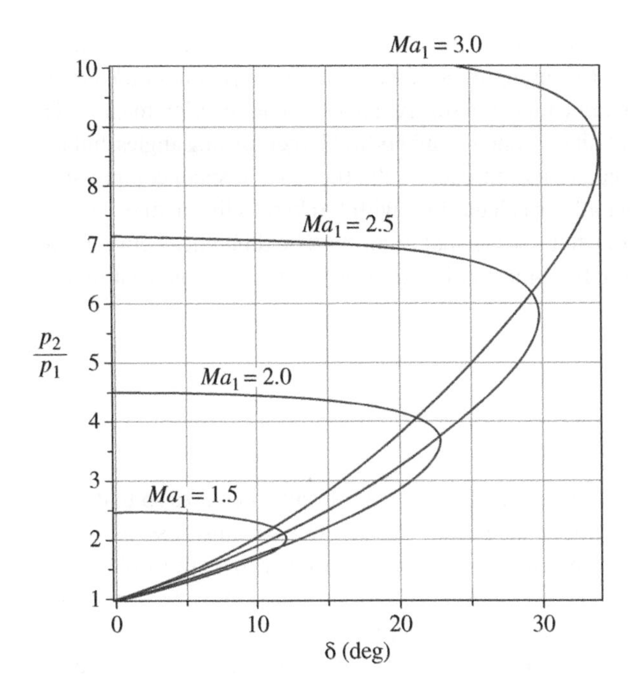


**Figure 8.6** Pressure after the shock ($p_2$) divided by pressure before the shock ($p_1$) as a function of deflection angles ($\delta$) for $\gamma = 1.4$. For external shocks use only the right-hand side of each *Ma*-curve up to the peak $\delta$-value. These curves are called "pressure-deflection shock polars." For larger range of parameters or better accuracy and other property values after the shock use the *Aerodynamics Calculator.*

$$\frac{p_2}{p_1} = \frac{2\gamma Ma_1^2 \sin^2\theta - (\gamma - 1)}{\gamma + 1} \tag{8.21}$$

$$\frac{T_2}{T_1} = \frac{\left[2\gamma Ma_1^2 \sin^2\theta - (\gamma - 1)\right]\left[(\gamma - 1)Ma_1^2 \sin^2\theta + 2\right]}{(\gamma + 1)^2 Ma_1^2 \sin^2\theta} \tag{8.22}$$

As previously stated, in external flows strong shocks occur when the flow meets a surface for which turning is so steep that no attached shock solution satisfies the governing equations. For such cases, a detached shock is produced and the flow accomplishes all needed turning within the subsonic portions. The topmost segment of each curve in Figure 8.6 at its left end where $\delta = 0°$ is the normal shock pressure jump for each respective incoming Mach number.

**Example 8.2** Be aware that this example is aimed at contrasting non-isentropic compressions like shocks with ideal-flow compressions. Using Figure 8.4 or the *Aerodynamics Calculator* for the Prandtl–Meyer function estimates the pressure *increase* in a compression turn for a flow at $M_1 = 3.0$ by an amount $\Delta\nu = -20°$.

For concave turning, we subtract the turning angle from that representing the incoming Mach number, namely, $\nu_2 = 49.76 - 20 = 29.76°$.

$$\nu_1 = 49.76° \quad \text{and} \quad p_1/p_t = 0.0272 \quad \text{at} \quad M_1 = 3.0$$

$$\nu_2 = 29.76° \quad \text{and} \quad p_2/p_t = 0.1052 \quad \text{at} \quad M_2 = 2.125$$

These results indicate an *isentropic pressure ratio* increase of $p_2/p_1 = 3.86$. The correct results from *oblique shock theory*, Eqs. (8.20) and (8.21) (using Figure 8.6 and/or the *Aerodynamics Calculator*) are $p_2/p_1 = 3.771$ with $M_2 = 1.99$, and this value reflects a somewhat lower compression magnitude.

We should expect larger discrepancies at higher Mach numbers and larger turning angles, but for lower values of turning angle the two calculations approach each other so that small compression angles can be close to isentropic. This fact allows calculations with "Ackeret's linear rule," which will be introduced in Chapter 9 in Eq. (9.7), to be somewhat accurate. The difference between the two given results reflects a deterioration of fluid-mechanical energy into heat that is what makes shocks irreversible.

## 8.6 Critical Mach Number

As air flows speed up over the front of an airfoil, the Mach number locally increases so that high enough compressibility effects may locally arise even in low subsonic free streams. Also, as supersonic aircraft accelerates to their cruising speed after take-off, Mach numbers over the airfoil transition from subsonic to transonic before reaching supersonic. These separate flow regimes must be properly distinguished because the character of the flow changes dramatically while transitioning from subsonic to supersonic. The transonic region is a mixture of both subsonic and supersonic types of flow – one that tends to produce large drag increases from shock formation and shock-induced flow separation. The *critical Mach number* is defined as the free stream Mach number at which sonic flow first appears over an airfoil. It may be found from the pressure coefficient defined earlier as Eq. (2.29), which is further manipulated here to become Eq. (8.24).

$$C_p = \frac{p - p_\infty}{q_\infty} = \frac{p_\infty}{q_\infty}\left(\frac{p/p_t}{p_\infty/p_t} - 1\right) \tag{8.23}$$

$$= \frac{p_\infty}{q_\infty}\left[\left(\frac{1 + \frac{\gamma-1}{2}Ma_\infty^2}{1 + \frac{\gamma-1}{2}Ma^2}\right)^{\frac{\gamma}{\gamma-1}} - 1\right]$$

$$\text{but} \quad \frac{\frac{p_\infty}{\rho_\infty}}{\frac{1}{2}V_\infty^2} = \frac{2RT_\infty}{V_\infty^2} = \frac{2}{\gamma Ma_\infty^2} \quad \text{hence}$$

$$C_p = \frac{2}{\gamma Ma_\infty^2}\left[\left(\left(\frac{1 + \frac{\gamma-1}{2}Ma_\infty^2}{1 + \frac{\gamma-1}{2}Ma^2}\right)^{\frac{\gamma}{\gamma-1}} - 1\right)\right] \tag{8.24}$$

In Eq. (8.23) we used the isentropic flow relation (Eq. (8.9)) to replace the pressure ratios after inserting the stagnation pressures.

Now to form the *critical Mach pressure coefficient* $C_{p,cr}$, we set $Ma = 1.0$ and change $Ma_\infty$ to $Ma_{crit}$ to symbolize the critical condition in Eq. (8.24). We then further associate $C_{p,cr}$ to the "Prandtl–Glauert relation" at this $Ma_{crit}$, which can be shown to represent the dependence of the pressure coefficient on incoming Mach number in subsonic flows. The Prandtl–Glauert relation formulates a pressure coefficient increase with Mach number; the origin of the Prandtl–Glauert relation will be discussed in Chapter 9 and identified as Eq. (9.4). $C_{p0}$ is the incompressible-flow value of the pressure coefficient. The *critical Mach number* $Ma_{crit}$ may be found from the solution of the middle and end portions of Eq. (8.25) and is plotted in Figure 8.7 for air – under critical Mach number conditions the pressure has a minimum value and the flow is sonic.

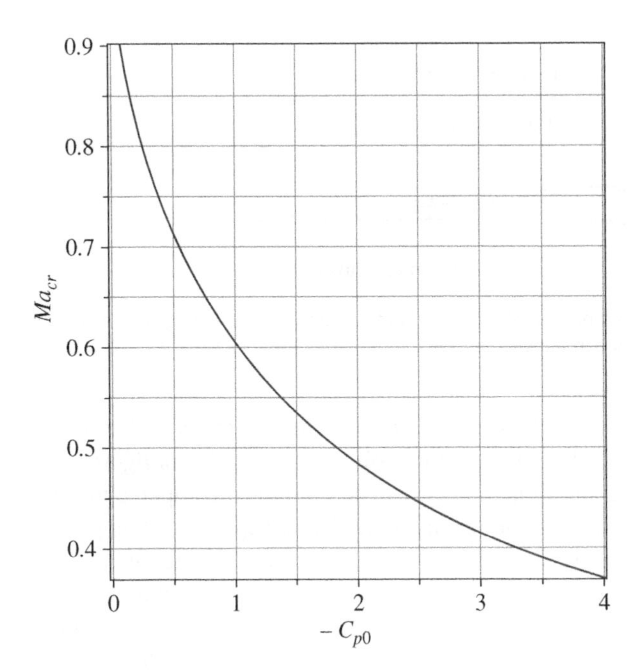


**Figure 8.7** Critical Mach number as given by the intersection of portions of Eq. (8.25). In the Prandtl–Glauert relation shown on the right of Eq. (8.25), values for $C_{p0}$ depend $\alpha$ and airfoil thickness and are negative.

$$C_{p,cr} = \frac{2}{\gamma\, Ma_{crit}^2}\left[\left(\frac{2+(\gamma-1)\,Ma_{crit}^2}{\gamma+1}\right)^{\frac{\gamma}{\gamma-1}} - 1\right] \equiv \frac{C_{p0}}{\sqrt{1-Ma_{crit}^2}} \tag{8.25}$$

For ordinary thin airfoils, resulting critical Mach numbers range between about 0.65 and 0.85. Using the incompressible-flow pressure coefficient ($C_{p0}$) allows representation of a variety of airfoils of interest when working with Eq. (8.25), where $C_{p0}$-values for airfoils are obtained either experimentally or analytically when $Ma < 0.3$.

In Figure 8.7, the value $C_{p0}$ used needs to be the lowest incompressible pressure coefficient, which is found at an airfoil's minimum pressure or highest velocity location. Recall that for any given airfoil configuration this minimum pressure location is also a function of $\alpha$. For incompressible flows around a cylinder, we found a maximum value of $C_{p0} = -3.0$ that, according to Figure 8.7, translates into a critical Mach number of about $Ma_{crit} = 0.42$ for circular cylinders in ideal flows. With thin symmetrical airfoils, Figure 5.5 shows higher values than −3.0 near the forward stagnation point, but these numbers arise from a mathematical singularity in the model. In other words, we should not assume that 3.0 represents an upper limit for $-C_{p0}$. In airfoils the highest magnitude of $C_{p0}$ is commonly found at the airfoil's "crest" (in symmetric ones this is at their thickest location but one that moves forward with angle of attack – although during the cruise condition where the evaluation of the critical Mach number is needed, angles of attack remain small). Real airfoils, therefore, do not quite go supersonic at their leading edge but rather at more central locations depending on $t_c/c$, where $t_c$ is the *airfoil's crest* that is near the maximum thickness $t_m$. Some advanced *high-lift airfoils* can exceed $C_{p0} = -3.0$ at their crests, and this makes possible $Ma_{crit}$-values under 0.4.

Flow pattern observations in wind tunnels indicate that when Mach numbers exceed the critical Mach number, a "supersonic flow bubble" develops over the airfoil that tends to terminate abruptly

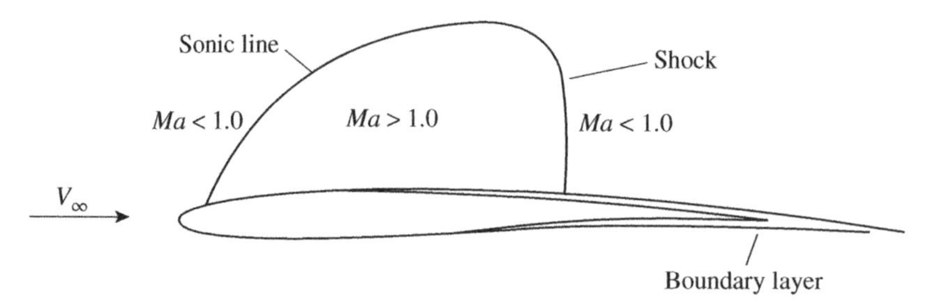


**Figure 8.8** Transonic flow showing the supersonic bubble that ends in a shock. [Additional images can be found by searching the web for "transonic shock images."]

in a nearly normal shock as shown in Figure 8.8. This shock often detaches the top boundary layer with a consequential large increase of form drag. Unless the airfoil is specifically configured for transonic flows, as discussed in Chapter 10, this drag increase can be substantial compared to skin drag. Such sudden drag rise was originally labeled the "sound barrier" and required adding power to the aircraft engines to overcome it. Presently, supersonic aircraft have, in addition to higher propulsive power, thin flat and relatively smaller wings as these can develop sufficient lift at high Mach numbers and produce less drag while traversing the transonic region. In Chapter 9 we continue with this topic and discuss swept wings, but here we turn next to the flat plate, the preferred fully supersonic airfoil configuration.

## 8.7 Supersonic Flat-plate Airfoils

Having discussed the two ways a supersonic flow can change direction, we examine next the simplest and most effective type of supersonic airfoil configuration, namely, the thin flat plate. As shown in Figure 8.9a, as the free stream flow approaches a two-dimensional flat plate at angle of attack, the need to turn the flow at the top surface (region 2) results in an expansion wave and at the lower surface (region 3) in an oblique shock. The flow after each of these waves is parallel to the flat plate, and within regions 2 and 3 it may be considered to be one-dimensional with respect to the plate; all this makes such configuration handy for using the equations developed in Section 8.4. In our analysis we proceed to establish sectional characteristics without consideration of boundary layers at either surface. Additionally, in Figure 8.9a, regions 4 and 5 indicate the flows at top and bottom leaving the plate's trailing edge parallel to each other (but not along the chord direction). Many downstream wave interactions have been omitted from Figure 8.9a as they do not affect the pressure forces at the airfoil's surface.

Pressure forces on the airfoil solely result from the set of waves preceding each location that gives rise to constant magnitude forces directed toward the surface as shown in Figure 8.9b. From this figure we can establish that the *net force vector* is directed up and to the right with a vertical component or *lift* and a horizontal component or *drag* – this drag is a purely supersonic effect and is called *wave drag*. Moreover, because of the symmetry of the pressure distributions, the aerodynamic center must now be located at the half-chord location instead of at quarter-chord as in subsonic flows.

In Example 8.3 we calculate resulting spanwise pressure changes together with the lift and drag forces on a flat-plate airfoil under supersonic flow, excluding boundary layers. These calculations

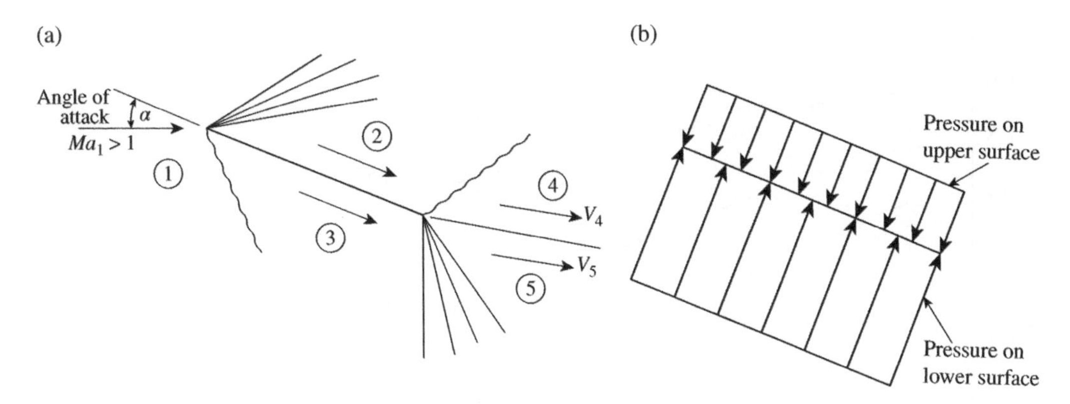


**Figure 8.9** (a) Two-dimensional flat-plate airfoil at angle of attack in a supersonic flow showing wave forms; (b) corresponding pressure distributions at the plate's surface with $p_3 > p_2$.

are then revisited in Chapter 9 to compare them with more approximate methods introduced there. Example 8.3 contains most of the ingredients relevant to two-dimensional supersonic airfoil calculations.

**Example 8.3** A flat-plate airfoil with a chord of 2 m is flying at an altitude of 4200 m ($p = 0.6$ bar or $0.6 \times 10^5$ N/m²) at $Ma_1 = 2.0$ at an angle of attack of 10°. Compute the lift and wave drag per unit span on this airfoil configuration. See Figure 8.8a for the relevant notation.

On the bottom surface, location 3 in Figure 8.6a, the pressure ratio through the oblique shock compression, using Figure 8.6, is $p_3/p_1 = 1.71$ so that $p_3 = 1.02 \times 10^5$ N/m². Now, on the top pressure, location 2, we can use Figure 8.4 before and after the Prandtl–Meyer expansion of $\Delta\nu = 10°$,

$$\nu_1 = 26.38° \quad \text{and} \quad p_1/p_t = 0.128 \quad \text{at} \quad Ma_1 = 2.0$$

$$\nu_2 = 26.38° + 10° = 36.38° \text{ so } p_2/p_t = 0.070 \text{ and } Ma_2 = 2.384$$

$$p_2 = [(p_2/p_t)/(p_1/p_t)] \times p_1 = 3.29 \times 10^4 \text{ N/m}^2$$

While Figures 8.4 and 8.6 are easy to use, they lack accuracy and working with the *Aerodynamics Calculator* is recommended. The lift and drag forces per unit span may now be found as [you should draw the applicable vector diagram]:

$$\ell = (p_3 - p_2) \times 2 \times \cos(10^\circ) = (1.02 - 0.329)(10^5)(2)(0.985) = 1.368 \times 10^5 \text{N/m}$$

$$dw = (p_3 - p_2) \times 2 \times \sin(10^\circ) = (1.02 - 0.329)(10^5)(2)(0.174) = 2.413 \times 10^4 \text{ N/m}$$

The drag $d_w$ is purely *wave drag* because we have not accounted for skin friction that is expected to contribute a comparable amount.

For ease of analysis, Figure 8.10 plots direct estimates of the static pressure ratio $p_2/p_1$ across the top expansion as a function of the approaching supersonic Mach number. In this chapter, Figure 8.10 along with its companion Figure 8.4 and Figure 8.6 (depicting the ratio of $p_3/p_1$ for oblique shocks) allow for quick pressure ratio estimates in regions 2 and 3 for the flat plate of Figure 8.9a.

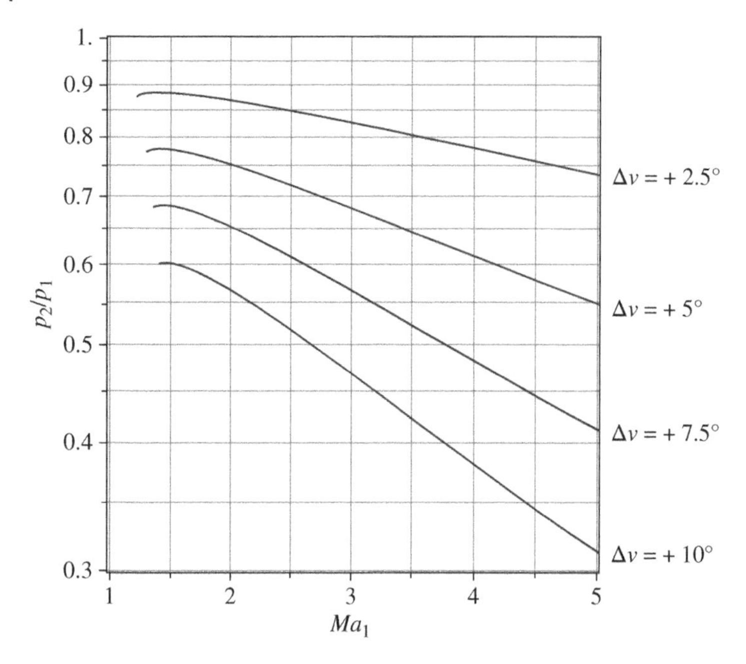


**Figure 8.10** Approximate values of expansion pressure ratios $p_2/p_1$ versus $Ma_1$ in air for $\Delta\nu$ = 2.5°, 5.0°, 7.5°, 10° in air for flows over a flat plate beyond the transonic range. For a larger range of parameters and better accuracy plus other property values after the expansion (such as $Ma_2$) use the *Aerodynamics Calculator.*

## 8.8 Enrichment Topic

### 8.8.1 Drag Under Supersonic Conditions

Compressibility changes the components of drag in significant ways. The portion of the drag represented by the skin-friction coefficient is known to decrease somewhat, becoming 60% at Mach 4 of its subsonic flow value. Wave drag, a strictly supersonic phenomenon, tends to contribute about the same amount as skin friction. In supersonic flows drag due to lift decreases because there is less need for high lift coefficients at supersonic Mach numbers and, depending on planform configuration with respect to free stream Mach angle, three-dimensional regions at the wings edges can be minimized as discussed in Chapter 9. Form or pressure drag, however, may be enhanced by the presence of shocks that tends to detach boundary layers in adverse pressure regions. Overall drag does increase with the onset of supersonic flows due primarily to the sum of skin friction and wave drag contributions, see Figure 8.11. A *drag divergence Mach number* is often assigned to the free stream Mach number value at which the drag rises abruptly, and this happens between the critical Mach number and the sonic or 1.0 value.

At a flat plate or any other sharp trailing edge, the issuing upper compression and lower expansion waves shown in Figure 8.9a merge in complicated ways to form the wake. In this wake both types of waves weaken as they interact by generating vorticity. In supersonic flows, therefore, vorticity plays an equivalent but more convoluted role than in subsonic flows.

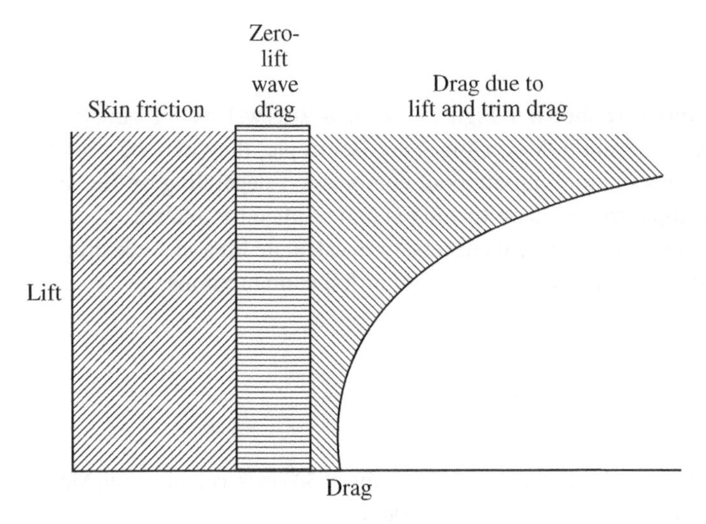


**Figure 8.11** Drag polar representation of relative magnitudes of total drag components in supersonic flows. *Source:* From Bertin and Cummings (2013).

## 8.9 Summary

The Mach number region where compressibility changes begin to affect the flow can be seen from Figures 8.2 and 8.7. Density ratios remain relatively unchanged up to $Ma = 0.3$ but start to drop noticeably beyond that. Temperature and pressure also drop with increasing Mach number at somewhat different rates, but it is the interaction between components in the momentum equation, first shown in incompressible form as Eq. (4.3) and in this chapter with Eqs. (8.2) through (8.6), that governs the interdependence of the pressure, temperature, and density on velocity. Equations (8.7)–(8.9) describe resulting isentropic compressible flow ratios as functions of $Ma$ and $\gamma$ and properly non-dimensionalized by existing stagnation values.

At supersonic conditions flow turning must involve either an oblique shock for concave turns or a Prandtl–Meyer expansion for convex turns because, as depicted in Figure 8.1a, gases moving supersonically cannot sense turns before arriving there. All oblique shocks, a requisite for compression turning, are inherently lossy and carry a stagnation or total pressure penalty, whereas supersonic expansion turns remain isentropic. For Prandtl–Meyer expansions the pressure at each Mach number can be found from Eq. (8.9) together with (8.12) and (8.13)

$$\frac{p}{p_t} = \left(\frac{T}{T_t}\right)^{\left(\frac{\gamma}{\gamma-1}\right)} = \left(1 + \frac{\gamma-1}{2}Ma^2\right)^{-\left(\frac{\gamma}{\gamma-1}\right)} \tag{8.9}$$

$$\nu = \left(\frac{\gamma+1}{\gamma-1}\right)^{1/2} \tan^{-1}\left[\frac{\gamma+1}{\gamma-1}(Ma^2-1)\right]^{1/2} - \tan^{-1}(Ma^2-1)^{1/2} \tag{8.12}$$

$$\nu_2 = \nu_1 + \Delta\nu \tag{8.13}$$

And for oblique shocks Eqs. (8.20) and (8.21) apply.

$$\tan\delta = 2(\cot\theta)\left(\frac{Ma_1^2\sin^2\theta - 1}{Ma_1^2(\gamma + \cos 2\theta) + 2}\right) \tag{8.20}$$

$$\frac{p_2}{p_1} = \frac{2\gamma Ma_1^2 \sin^2\theta - (\gamma - 1)}{\gamma + 1} \tag{8.21}$$

Relevant pressure ratios during turns are shown in Figures 8.4 and 8.10 as well as in Figure 8.6 as a function of Mach number. Because Prandtl–Meyer turns are isentropic, they may be used to describe either supersonic expansions or small amounts of compression depending on the sign of $\Delta\nu$ in Eq. (8.13), although isentropic compression turning is more difficult to realize.

Flat plates at angle of attack exhibit the ingredients necessary to calculate lift and wave drag in supersonic flows around thin and symmetric two-dimensional airfoil configurations.

## Problems

**8.1** For a perfect gas with $R = 2080$ N-m/kg-K and $\gamma = 1.67$ flowing at 500 m/s, calculate the Mach number and the stagnation pressure ratio when $T = 1000$ K.

**8.2** For the gas in Problem 8.1 calculate the stagnation pressure coefficient ($C_{pt}$) and, if this were the critical Mach number, the critical pressure coefficient ($C_{p,cr}$).

**8.3** Calculate the magnitude of the force vectors per unit span in sections 1, 2, and 3 and the net lift and wave drag on the entire airfoil section. The orientation of surface 3 is 4° above the velocity vector and the approaching Mach number is 3.0. Take the chord length as 2 m and the ambient pressure as 0.592 atm. The isosceles triangular section has 8° equal angles.

**Figure P8.3**

**8.4** A symmetrical diamond-shaped airfoil is shown in Figure P8.4.
a) Show the waves preceding each surface as well as those at the tail.
b) At what angle of attack will there be no compression waves on the first upper surface?
c) Calculate the total lift and drag and drag on the airfoil section with the shown information.

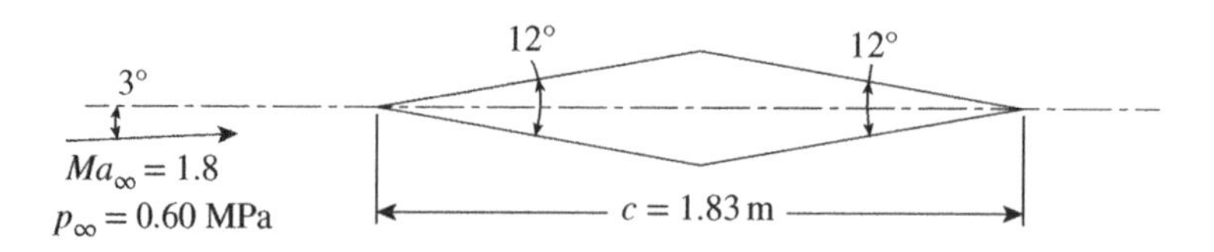


**Figure P8.4**

**8.5** It turns out that airfoils of any thickness in supersonic flow will develop wave drag even at zero lift. To show this calculate the wave drag per unit span when the air flow is at $Ma_1 = 3.0$ and $\alpha = 0°$ on the symmetric object in Figure P8.5. The chord length is 5 m and the atmospheric altitude is 14.4 km. The total end angle of diamond-shaped airfoil is 6°.

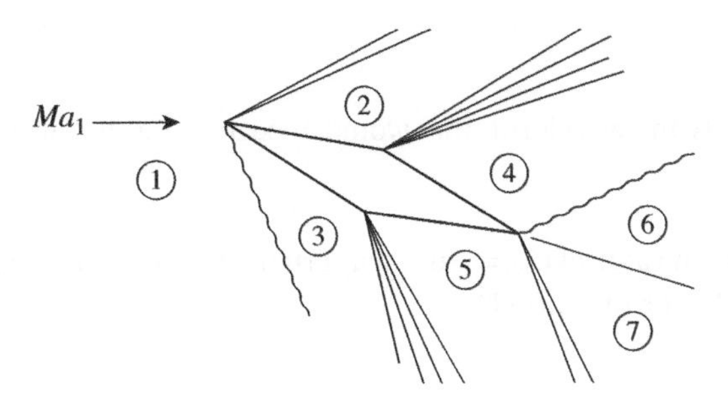


**Figure P8.5**

## Check Test

**8.1** Explain in your own words why supersonic expansions are isentropic while supersonic compressions are nearly always not.

**8.2** Discuss why supersonic flows need oblique shocks or Prandtl–Meyer expansions to make turns.

**8.3** Why is the standard form of Bernoulli's equation introduced in Chapter 4 not appropriate in Chapter 8?

**8.4** What is the difference between a normal shock and an oblique shock with respect to the flow meeting boundary conditions?

**8.5** Under what geometrical conditions may we treat the flow before and after an oblique shock or a centered expansion as one-dimensional?

**8.6** Write the values of the Prandtl–Meyer angle $\nu$ and the Mach angle $\mu$ at $Ma = 1.0$ and at 5.0. Discuss their expected trends with increasing Mach number.

**8.7** Why does a thin flat plate represent the best sectional shape for a supersonic airfoil?

**8.8** Explain the pressure contributions to the lift force in supersonic airfoils in terms of oblique shocks and Prandtl–Meyer turns.

**8.9** Under what condition would the forward wave at surface 2 disappear in Figure P8.5?

**8.10** What is the specific feature of supersonic flows that give rise to wave drag?

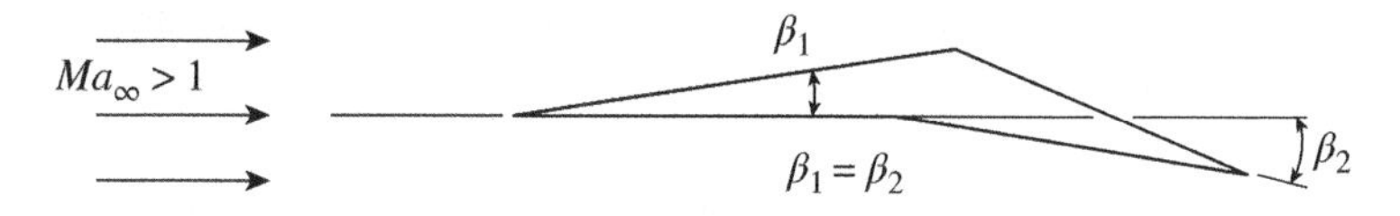


**Figure CT8.10**

**8.11** In Figure CT8.10, sketch and identify the waveforms that would appear on the airfoil shape depicted.

**8.12** An NACA 0012 airfoil has a minimum value of $C_{po} = -0.4$ at a certain angle of attack. What would the Critical Mach Number be for this airfoil?

# 9

# Thin Airfoils in Compressible Flow

## 9.1 Introduction

We continue our study of thin airfoils by examining ideal-compressible flows in two Mach number regimes. The first comprises purely subsonic Mach numbers ranging from 0.3 to about 0.8 and the second entirely supersonic starting from above Mach 1 up to Mach 5. Transonic and hypersonic flows are covered in Chapter 10. In Chapter 8 we present the necessary background to calculate sectional lift and wave drag forces on a flat plate under supersonic flow. In this chapter we introduce an important set of thin airfoil compressible flow *similarity rules* based on the linearized small perturbation equation for the potential function (first introduced in Chapter 2 as Eq. (2.38)) to analyze such flows – the advantage of this approach is its relative simplicity since supersonic compressions are treated as reversible. In our subsonic formulation we make use of the incompressible-flow pressure coefficient ($Cp_0$) from Chapter 5 as well as of the isentropic flow relations from Chapter 8. We also introduce in Chapter 8 the *critical Mach number* ($Ma_{cr}$) to identify purely subsonic flow ranges over an airfoil. Moreover, we have uncovered a strictly supersonic contribution to the drag – *wave drag*. In this chapter we will discuss why airfoil camber and thickness do not enhance the lift but do increase the drag of supersonic airfoils. We will also introduce the reasoning behind wing *sweepback*. A set of modified aerodynamic coefficients more appropriate than those presented in Chapter 2 for studying the effects of Mach number on lift and pitching moments are developed in this chapter.

## 9.2 Objectives

After completing this chapter successfully, you should be able to:

1) Describe under what condition the small perturbation equation for the velocity potential becomes *Laplace's equation* and give some general properties of its solution.
2) Describe under what condition the small perturbation equation becomes the *wave equation* and contrast its mathematical character to *Laplace's equation*.
3) State the *Prandtl–Glauert rule* for purely subsonic flows and its significance and limitations.
4) State the *Ackeret rule* for purely supersonic flows over a flat plate and its significance and limitations.
5) Give the two-dimensional airfoil section location of the *aerodynamic center* in both subsonic and supersonic flows. Explain the reason for the difference.
6) Discuss why the *lift* itself continuously increases with Mach number outside of the transonic region even though the lift-coefficient slope from *Ackeret's rule* decreases in the supersonic

*Elements of Aerodynamics: A Concise Introduction to Physical Concepts*, First Edition. Oscar Biblarz.

Companion website: www.wiley.com/go/elementsofaerodynamics

region. Infer from this why wings flying at compressible Mach numbers needs less planform area and/or less angle of attack to achieve the same lift.

7) Discuss why *wave drag* is strictly a supersonic effect, and why increases of this wave drag render camber and thickness undesirable for supersonic flight.
8) Give reasons why *delta wings* or *straight stubby wings* are used in high supersonic flow in terms of the critical Mach number and of wing regions of two- and three-dimensional flow.
9) Describe the advantage of *swept wings* in subsonic flow in terms of the angle of sweep ($\Lambda$) and of the approaching flow Mach angle ($\mu_\infty$).

## 9.3 Two-Dimensional Compressible Flow Around Thin Airfoils

We will refer here to some well-known properties of both Laplace's and the wave equation that relate to our default steady-state flow model (recall that for our reference frame we "jump on an airfoil moving at a constant velocity $V_\infty$" so pressure distributions and supersonic wave fronts appear stationary). In this section we introduce the Prandtl–Glauert (subsonic) and Ackeret (supersonic) rules because for our Mach number domains of interest they allow convenient interpretations of Mach number effects based on the $\left(1-Ma_\infty^2\right)$ term in Eq. (9.1). Previously introduced in Chapter 2 as Eq. (2.38), equation (9.1) describes irrotational supersonic flows derived from small flow disturbances around thin airfoils. Here $y$ is the cross-flow coordinate in two dimensions.

$$\left(1-Ma_\infty^2\right)\frac{\partial^2\phi}{\partial x^2}+\frac{\partial^2\phi}{\partial y^2}=0 \tag{9.1}$$

This small perturbation equation has a number of useful properties. When $Ma_\infty < 1.0$ this equation is elliptic (e.g. Laplace's equation), whereas if $Ma_\infty > 1.0$ it is hyperbolic (e.g. the wave equation). Subsonic flows may be studied with *harmonic* functions, whereas supersonic flows are represented by *stationary waves*, like those seen in Figure 8.1 where time is not explicit, and zones of action and zones of silence exist within appropriate Mach cone regions. Equation (9.1) has been extensively studied and documented; our task here is to decompose its mathematical nature in order to examine results for two-dimensional thin airfoils. Figure 9.1 depicts the pressure coefficient $C_p$ in purely subsonic and supersonic flow regimes around a thin airfoil.

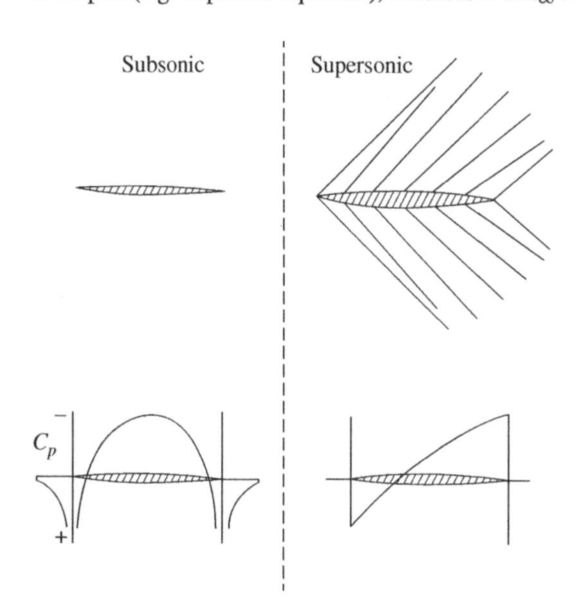


**Figure 9.1** Pressure coefficient in subsonic and supersonic flow regimes. *Source:* Adapted from Liepmann and Roshko (1957).

Harmonic functions are used in solving a variety of subsonic flow profiles through the well-known properties of the trigonometric functions that constitute the *Fourier series*. Under subsonic flow conditions, the largest magnitude of any variable must occur at the airfoil's edge with the free stream. In supersonic flows all disturbances propagate along the direction of the *characteristics* or downstream running *Mach lines* since when each functional argument in Eq. (9.2) shown next remains constant, the dependent variable $\emptyset$ must remain constant. The notation means f (...) for *upper* and g(...) for *lower* characteristic waves.

$$\emptyset(x,y) = \underset{\text{(upper)}}{f\left(x - \left(\sqrt{Ma_\infty^2 - 1}\right)y\right)} + \underset{\text{(lower)}}{g\left(x - \left(\sqrt{Ma_\infty^2 - 1}\right)y\right)} \tag{9.2}$$

One particularly useful result from *small perturbation theory* is given as Eq. (9.3). In this equation $u$ is the small perturbation velocity in the $x$-direction and here only linear terms have been retained. This pressure coefficient form applies in purely subsonic or purely supersonic flow domains and not in the transonic or hypersonic regions. Equation (9.3) together with the equations introduced in Chapter 2 for the lift, drag, and moment coefficients in terms of the pressure coefficient will be used to establish the formulations in this chapter.

$$C_p = \frac{p - p_\infty}{\frac{1}{2}\rho_\infty V_\infty^2} = -\frac{2u}{V_\infty} = -\frac{2}{V_\infty}\frac{\partial\phi}{\partial x} \tag{9.3}$$

### 9.3.1 Purely Subsonic Flows

The study of effects of compressibility aims at understanding observed increases on the pressure coefficient with Mach number. Equation (9.4) is perhaps the simplest analytical result for this, called the *Prandtl–Glauert rule* after its originators, where $C_{p0}$ is the incompressible value of the pressure coefficient. In Section 9.6 we motivate the reasoning behind the form of Eq. (9.4). Results from this rule slightly underpredict data though such accuracy is compatible with other assumptions. Equation (9.5b) is the resulting integral for the two-dimensional lift coefficient writing the lower minus the upper values as $(C_{p,l} - C_{p.u})_0$ and Eq. (9.6) is the associated moment coefficient. Equation (9.5b) is the slope of the lift coefficient $(dc_\ell/d\alpha)$ using $c_{\ell 0} = 2\pi\alpha$ in ideal-incompressible flows. It can also be shown that the aerodynamic center remains at the quarter chord $(c/4)$. Because the factor $\sqrt{1 - Ma_\infty^2}$ appears in the denominator of the following equations, we surmise that these coefficients grow in magnitude with $Ma_\infty$-increases. This growth accelerates beyond $Ma_\infty = 0.6$ leading to an infinite value at Mach 1. For this and other reasons we cannot, therefore, use the Prandtl–Glauert rule in the transonic region, and this is further elaborated upon in Section 9.4.

$$C_p = \frac{C_{p0}}{\sqrt{1 - Ma_\infty^2}} \tag{9.4}$$

$$c_\ell = \frac{1}{\sqrt{1 - Ma_\infty^2}}\frac{1}{c}\int_0^c \left(C_{p,l} - C_{p.u}\right)_0 dx = \frac{c_{\ell 0}}{\sqrt{1 - Ma_\infty^2}} \tag{9.5a}$$

$$c_{\ell\alpha} = \frac{c_{\ell\alpha 0}}{\sqrt{1 - Ma_\infty^2}} = \frac{2\pi}{\sqrt{1 - Ma_\infty^2}} \tag{9.5b}$$

$$c_m = \frac{c_{m0}}{\sqrt{1 - Ma_\infty^2}} \tag{9.6}$$

**Example 9.1** Incompressible-flow measurements on an NACA 2412 airfoil indicate that $c_{\ell 0} = 0.78$ at $\alpha = 6°$. Using the Prandtl–Glauert rule, calculate the resulting values of the lift coefficient at $Ma_\infty = 0.3$, 0.6, and 0.9.

Using Eq. (9.5a) we find that $c_\ell = 0.82$ at $Ma_\infty = 0.3$, 0.975 at $Ma_\infty = 0.6$, and $c_\ell = 4.10$ at $Ma_\infty = 0.9$. The first result is only slightly greater that the incompressible value, but the curve rapidly increases, with the last result having a value five times higher. While this growth in lift coefficient has the proper trend with increasing Mach number, Eq. (9.5a) cannot be applied for values of $Ma_\infty > 0.7$ (in the transonic flow range).

### 9.3.2 Purely Supersonic Flow Over Two-Dimensional Flat Plates

Thin flat plates are the best theoretical model for such airfoils because in supersonic flow both thickness and camber generate the *wave drag* as discussed in Section 8.7 (see Figure 8.9b). Moreover, for supersonic flow over flat plates the center of pressure and aerodynamic center coincide. The supersonic counterpart to the Prandtl–Glauert rule is *Ackeret's rule* shown as Eqs. (9.7), (9.8a), (9.8b), and (9.9). In Eq. (9.7) the plus sign applies to the lower plate surface and the minus sign to the upper surface at angle of attack ($\alpha$).

$$C_p = \frac{\pm 2\alpha}{\sqrt{Ma_\infty^2 - 1}} \tag{9.7}$$

$$c_\ell = \frac{4\alpha}{\sqrt{Ma_\infty^2 - 1}} \tag{9.8a}$$

$$c_{\ell\alpha} = \frac{4}{\sqrt{Ma_\infty^2 - 1}} \tag{9.8b}$$

$$c_{dw} = \frac{4\alpha^2}{\sqrt{Ma_\infty^2 - 1}} \tag{9.9}$$

The slope of the lift coefficient for a flat plate is given in Ackeret's rule as Eq. (9.8b). In Eq. (9.9), $c_{dw}$ is the wave drag acting on the thin flat plate that adds to frictional effects. Contributions of camber and thickness may be calculated for wedge profiles with the oblique shock and Prandtl–Meyer expansion formulas introduced in Chapter 8 as is shown in Section 9.5.

We have already seen in Example 8.3 that wave drag arises on a lifting flat plate from pressure forces that act perpendicular at the surface giving rise to Eq. (9.9). In curved airfoil surfaces the force vectors are distributed and weak oblique shocks are generated to fit the boundary conditions – actual details are beyond the scope of this book.

### 9.3.3 Supersonic Flow Moment Coefficient

As shown in Chapter 8, the aerodynamic center in *supersonic symmetric airfoils* is at the half-chord location ($c/2$) due to the symmetry of the pressure distributions that makes the pitching moment about this aerodynamic center zero. Under thin airfoil theory, the pitching moment at any arbitrary chord location $x_0$ may be written as

$$c_{mx_0} = \frac{-4\alpha}{\sqrt{Ma_\infty^2 - 1}}\left(\frac{1}{2} - \frac{x_0}{c}\right) \tag{9.10}$$

**Example 9.2** Compare the results from Example 8.3 with calculations from Eqs. (9.8a) and (9.9). In Example 8.3 we looked at a flat plate airfoil with $c = 2$ m is flying at an altitude of 4200 m ($p_\infty = 0.6$ bar) at $Ma_\infty = 2.0$ with an angle of attack of 10°.

The values calculated in the example of the previous chapter were $\ell = 1.368 \times 10^5$ N/m and $d_w = 2.413 \times 10^4$ N/m. Since for Example 8.3 $q_\infty = \left(\frac{\gamma}{2}\right) p_\infty M_{a\infty}^2 = 1.68 \times 10^5$, we get $c_\ell = 0.407$ and $c_{dw} = 0.0715$. The *coefficients* as calculated with Eqs. (9.8a) and (9.9) become $c_\ell = 0.403$ and $c_{dw} = 0.07035$ or around 2% lower.

It is expected that such differences would further increase at higher angles of attack because shock strengths increase. Due to the constancy of the pressure distributions above and below the flat plate, both the center of pressure and the aerodynamic center coincide and the pitching moment about them is zero. For airfoils with *curved surfaces*, the application of small perturbation theory though less accurate can be much less cumbersome.

## 9.4 The Mach Number Dependence

In order to properly study the Mach number dependence of the aerodynamic coefficients in compressible flow, we need to modify the definition of the lift and moment coefficients as given in Chapter 2. This is because the dynamic pressure ($q_\infty$), which contains the free stream Mach number, appears in the denominator to make the pressures non-dimensional. In Chapter 8 we presented a form for the pressure coefficient in isentropic flows in Eq. (8.24) that involves only Mach numbers and $\gamma$. The Prandtl–Glauert and Ackeret rules are based on weak flow-perturbing effects around the airfoil so that compressions are not far from reversible (as already shown in Examples 8.2 and 9.2) so that the isentropic nature implicit in Eqs. (9.1) and (9.3) is relevant. In other words, supersonic compressions are treated as *inverted Prandtl–Meyer expansions* in Ackeret's rule but not as accurately as with the oblique shock formulations.

The subsonic compressible sectional *lift-slope coefficient* in the Prandtl–Glauert rule tends to rapidly increase with Mach number from its initial incompressible value up to the transonic region (as shown in Example 9.1). On the supersonic side, the lift-slope coefficient has a comparably rapid slope decrease that, at first glance, might suggest limiting supersonic flights to low Mach numbers to maintain the lift coefficients at reasonable values (see Problem 9.1). However, these trends are deceiving as the lift force itself behaves differently when the Mach number does not appear in the denominator of $c_\ell$ for two reasons:

1) There can be no lift at $Ma_\infty = 0$ in spite of what Eq. (9.5a) might infer.
2) The magnitude of the lift force continuously increases with Mach number outside of the transonic region contrary to the trend seen in Eq. (9.8a).

In the Tables 9.1, 9.2, and 9.3 we introduce a set of *modified dimensionless coefficients* that do not contain the Mach number, using the new symbol $\xi$ to replace the coefficient $c$. In order to eliminate $Ma_\infty$ and retain our newly defined coefficients in dimensionless form, make use of a relation equivalent to $\frac{1}{2}\rho_\infty V_\infty^2$, namely,

$$q_\infty = \left(\frac{\gamma}{2}\right) p_\infty Ma_\infty^2$$

**Table 9.1** Modified coefficient definitions.

| Conventional | Modified |
|---|---|
| $C_p = \dfrac{p - p_\infty}{q_\infty}$ | $\xi_p \equiv \dfrac{2(p - p_\infty)}{\gamma p_\infty}$ |
| $c_\ell = \dfrac{\ell}{q_\infty c}$ | $\xi_\ell \equiv \dfrac{2\ell}{\gamma p_\infty c}$ |
| $c_m = \dfrac{m}{q_\infty c^2}$ | $\xi_m \equiv \dfrac{2m}{\gamma p_\infty c^2}$ |
| $c_{dw} = \dfrac{d_w}{q_\infty c}$ | $\xi_{dw} \equiv \dfrac{2d_w}{\gamma p_\infty c}$ |

**Table 9.2** Subsonic compressible sectional coefficients.

| Conventional | Modified |
|---|---|
| $C_p = \dfrac{C_{p0}}{\sqrt{1 - Ma_\infty^2}}$ | $\xi_p \equiv \dfrac{C_{p0} Ma_\infty^2}{\sqrt{1 - Ma_\infty^2}}$ |
| $c_\ell = \dfrac{2\pi(\alpha - \alpha_{\ell 0})}{\sqrt{1 - Ma_\infty^2}}$ | $\xi_\ell \equiv \dfrac{2\pi(\alpha - \alpha_{\ell 0}) Ma_\infty^2}{\sqrt{1 - Ma_\infty^2}}$ |
| $c_m = \dfrac{c_{m0}}{\sqrt{1 - Ma_\infty^2}}$ | $\xi_m \equiv \dfrac{c_{m0} Ma_\infty^2}{\sqrt{1 - Ma_\infty^2}}$ |

**Table 9.3** Supersonic flat-plate sectional coefficients.

| Conventional | Modified |
|---|---|
| $C_p = \dfrac{2\alpha}{\sqrt{Ma_\infty^2 - 1}}$ | $\xi_p \equiv \dfrac{2\alpha Ma_\infty^2}{\sqrt{Ma_\infty^2 - 1}}$ |
| $c_\ell = \dfrac{4\alpha}{\sqrt{Ma_\infty^2 - 1}}$ | $\xi_\ell \equiv \dfrac{4\alpha Ma_\infty^2}{\sqrt{Ma_\infty^2 - 1}}$ |
| $c_{m0} = \dfrac{2\alpha}{\sqrt{Ma_\infty^2 - 1}}$ | $\xi_{m0} \equiv \dfrac{2\alpha Ma_\infty^2}{\sqrt{1 - Ma_\infty^2}}$ |
| $c_{dw} = \dfrac{4\alpha^2}{\sqrt{Ma_\infty^2 - 1}}$ | $\xi_{dw} \equiv \dfrac{4\alpha^2 Ma_\infty^2}{\sqrt{Ma_\infty^2 - 1}}$ |

Table 9.1 shows our modified coefficient definitions made non-dimensional by $p_\infty$ multiplied by the unitless constant $\gamma/2$. Table 9.2 develops the resulting two-dimensional forms for subsonic thin airfoils and Table 9.3 for flat plates in supersonic flow. Any use of the Prandtl–Glauert and Ackeret rules implies small angles of attack in both flow regimes. [You should be able to verify that the coefficients introduced next are indeed non-dimensional.]

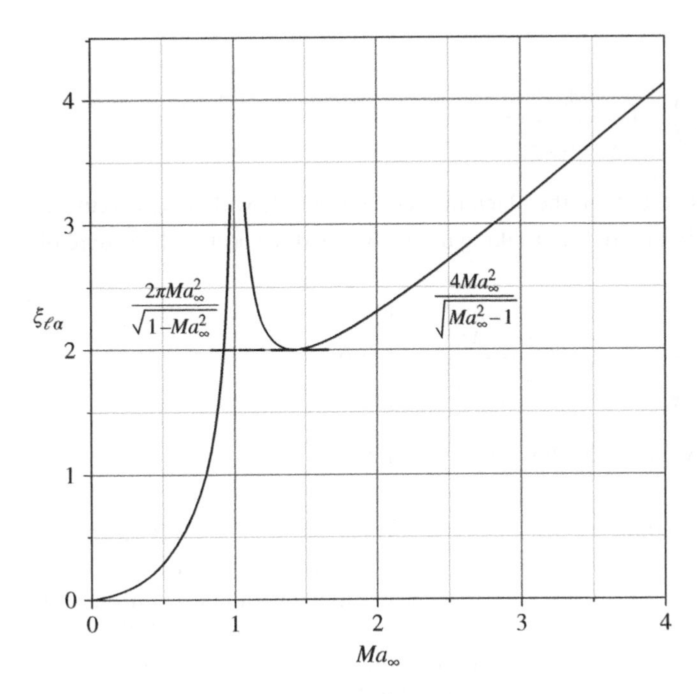


**Figure 9.2** Plot of the two-dimensional *lift-slope factor* $\xi_{\ell\alpha} \equiv c_{\ell\alpha} Ma_\infty^2 = d(2\ell/\gamma p_\infty c)/d\alpha$ based on *small perturbation theory* results. Solid curves between $Ma_\infty = 0.8$ and 1.2 are in the transonic region and display unrealistic trends. Beyond Mach = 4 the dependence of $\xi_{\ell\alpha}$ becomes linear with $Ma_\infty$.

We will continue by further examining the behavior of the *slope of the lift coefficient* in both subsonic and supersonic flows,

$$\xi_{\ell\alpha} \equiv c_{\ell\alpha} Ma_\infty^2$$

as a function of the incoming Mach number because this coefficient depends on the fewest parameters. Figure 9.2 shows the *lift-slope factor* ($\xi_{\ell\alpha}$) plotted as a function of approaching Mach number. Note that outside of the transonic region our non-dimensional slope $\xi_{\ell\alpha}$ always increases with $Ma_\infty$ as shown by the curves bridged by the dashed line; experiments show the transonic region is crossed with slopes of finite magnitude but of a more complicated lift pattern. Bridging the transonic region this way avoids the singularities inherent with both the Prandtl–Glauert and Ackeret rules.

Compressibility manifests itself in supersonic flows by decreasing the conventional pressure coefficient with increasing Mach numbers with a similar effect on the other coefficients that derive from it. However, as can be seen in Figure 9.2, at $Ma_\infty = \sqrt{2}$(or 1.414) our modified *lift* coefficient is seen to go through a *minimum*. The conventional *lift-slope coefficient* ($c_{\ell\alpha}$) has the lift divided by the *dynamic pressure* that contains the Mach number squared. In our more appropriate lift-slope factor $\xi_{\ell\alpha}$, this minimum appears just beyond transonic region. The sectional lift itself continuously increases with Mach number in flat plates operating at $Ma_\infty > \sqrt{2}$. This minimum point is the subject of Example 9.3.

**(Optional) Example 9.3** Show that the two-dimensional lift slope (from Eq. (9.8a)) for a flat plate in supersonic flow has a minimum at $Ma_\infty = \sqrt{2} = 1.414$.

$$\ell = \frac{4\alpha}{\sqrt{Ma_\infty^2 - 1}}(q_\infty c) \quad \text{and} \quad q_\infty = \frac{1}{2}\rho V_\infty^2 = \frac{1}{2}\gamma p_\infty Ma_\infty^2$$

The lift may then be written as

$$\ell = \frac{4\alpha c}{\sqrt{Ma_\infty^2 - 1}}\left(\frac{1}{2}\gamma p_\infty Ma_\infty^2\right) = k\frac{Ma_\infty^2}{\sqrt{Ma_\infty^2 - 1}}$$

where $k \equiv 2\alpha\gamma p_\infty c$ to include terms other than the Mach number. We can show that $\ell_{min}$ occurs at $Ma_\infty = \sqrt{2}$ either by plotting $\xi$ as in Figure 9.4 or taking derivatives with respect to the square of the Mach number, i.e.

$$\frac{d}{dMa_\infty^2}\left[k\frac{Ma_\infty^2}{\sqrt{Ma_\infty^2 - 1}}\right] = k\left[\frac{1}{\sqrt{Ma_\infty^2 - 1}} - \frac{1}{2}\frac{Ma_\infty^2}{\sqrt[3]{Ma_\infty^2 - 1}}\right] = 0$$

$Ma_\infty^2 - 1 = \frac{1}{2}Ma_\infty^2$ or $Ma_\infty^2 = 2$ which is the lift-curve minimum.

**Example 9.4** Arrive at equations required to calculate planform areas for an aircraft capable of varying its wing geometry so as to cruise at two subsonic Mach numbers, say at $Ma_{\infty 1} = 0.3$ and $Ma_{\infty 2} = 0.7$. Assume the aircraft has a built-in mechanism to change its wing configuration and that it weights the same in both cases. We will assume further that the wing performs with the same efficiency at all planform values, with proportional two-dimensional and three-dimensional coefficients and $c_{\ell 0} = a\alpha$ (incompressible three-dimensional slope, Eq. (6.17)).

We should expect that the larger planform area is used at the lower Mach number. Under the stated simplifications and since the weights are equal, we need only equate lift force components at the two given flight Mach numbers. From Eq. (9.5a) equating the total lift forces:

$$\frac{a_2\alpha_2\left(\frac{\gamma}{2}\right)p_{\infty 2}Ma_{\infty 2}^2 S_2}{\sqrt{1 - Ma_{\infty 2}^2}} = \frac{a_1\alpha_1\left(\frac{\gamma}{2}\right)p_{\infty 1}Ma_{\infty 1}^2 S_1}{\sqrt{1 - Ma_{\infty 1}^2}}$$

$$\frac{a_2\alpha_2 p_{\infty 2}S_2}{a_1\alpha_1 p_{\infty 1}S_1} = \frac{\sqrt{1 - Ma_{\infty 2}^2}}{\sqrt{1 - Ma_{\infty 1}^2}}\frac{Ma_{\infty 1}^2}{Ma_{\infty 2}^2} = 0.138$$

Supposing now that planform changes are made with a "telescoping wing" that doubles the planform area by extending the span dimension without significantly changing its incompressible three-dimensional slopes (i.e. $a_2/a_1 \sim 1.0$ thus focusing only on compressibility effects), and assuming that the low speed flight is at sea level whereas the high speed cruise is at 10,200 m (where $p_{\infty 2} = 2.57 \times 10^4\ \text{N/m}^2$). The resulting ratio in angles of attack becomes:

$$\frac{\alpha_2}{\alpha_1} = 0.138\frac{p_{\infty 1}S_1}{p_{\infty 2}S_2} = 0.138 \times \frac{10.13}{2.57} \times 2.0 = 1.084$$

Under the given conditions the needed change of angle of attack would be insignificant. This example, however, shows that only 50% of the incompressible wing planform area is required at the higher Mach number with the angle of attack $\alpha$ remaining relatively unchanged. Such contrast gets even higher at supersonic speeds.

In reality aircraft can vary $\alpha$ more easily than undergo the more complicated adjustments for such a large change in the planform area (not to mention the sophisticated and heavy structural components necessary). Variable planform area aircraft remains somewhat rare, but they can be a good option for lowering the enhanced *drag divergence* that significantly governs engine power requirements as Mach numbers approach unity.

## 9.5 Supersonic Airfoils

In order to demonstrate that the best two-dimensional supersonic airfoil configurations are thin symmetrical sections, we will extend earlier results for a flat plate in Example 8.3 by making numerical comparisons with two other airfoil sections that have "thickness," one cambered and the other symmetric. As already stated, when supersonic flows first meet a "blunt object," such as the nose of our default teardrop-shaped subsonic airfoil, a *detached normal shock* forms wherein the flow can readily turn and somewhere after the nose this normal shock usually bends to form an oblique shock. Normal shocks are quite undesirable because of the high *form drag* they produce. The front end of supersonic airfoil therefore needs to be pointed. At the airfoil's thin aft-ends during supersonic flight, waves must form to locally equalize upper and lower pressures but not necessarily to meet the Kutta condition.

In order to apply the calculational schemes from Chapter 8, we examine flows over flat airfoil surface sectors in (i) an isosceles triangle as a model for a cambered airfoil (because its midpoint falls above the chord location) and (ii) a diamond-shaped (or double-wedge profile) airfoil as a model for a thick but symmetrical airfoil. These airfoil shapes have regions of one-dimensional flow away from their corner waves and can be analyzed with the more accurate formulations developed in Chapter 8. In Example 8.3 the incoming Mach number is 2.0. Our desired new shapes are shown in Figures 9.3a and b where the airfoil's included angles at each end are 5° in case (a) and 10° in case (b) – the result of having folded the first triangle about its chord line. At an angle of attack of 10°, each airfoil will first experience a Prandtl–Meyer expansion at the top and an oblique shock at the bottom but of different strengths compared to those in Example 8.3. The remaining waves and resulting pressure ratios will also differ but are not shown and are left as an exercise to the reader.

For each configuration $Ma_\infty = 2.0$ and $\alpha = 10°$. Results from calculations are shown as follows:

| | | |
|---|---|---|
| Configuration (a) | $\ell = 1.334 \times 10^5$ N/m $d_w = 2.55 \times 10^4$ N/m | Camber and thickness |
| Configuration (b) | $\ell = 1.294 \times 10^5$ N/m $d_w = 5.14 \times 10^4$ N/m | Thickness |

The lift coefficients for the airfoil sections of Figure 9.3 are somewhat lower compared to Example 8.3 where the lift was $\ell = 1.368 \times 10^5$ N/m. Both drag coefficients show increases compared to $d_w = 2.413 \times 10^4$ N/m in Example 8.3. This means that thickness and camber only contribute undesirable effects. The biggest discrepancy between the two new drag values stems from the oblique shock, which is stronger in configuration (b) causing larger losses. The input numbers used here contain only small angles and hence the flow may remain close to isentropic. [You need to know how to work Problems 8.3 and 8.4 in Chapter 8 to understand details related to the calculations.]

### 9.5.1 Rectangular Wings in Supersonic Flow

When flying supersonically rectangular wings of high-aspect ratio may contain substantial two-dimensional flow regions away from the wing tips. The extent of these zones depends directly

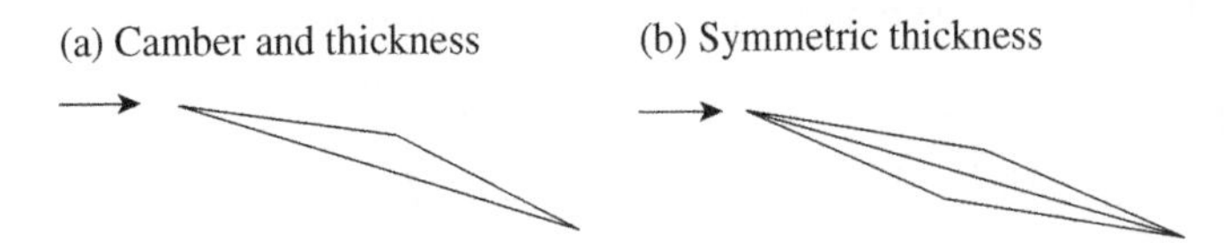


**Figure 9.3** (a) Triangular and (b) diamond airfoil cross sections.

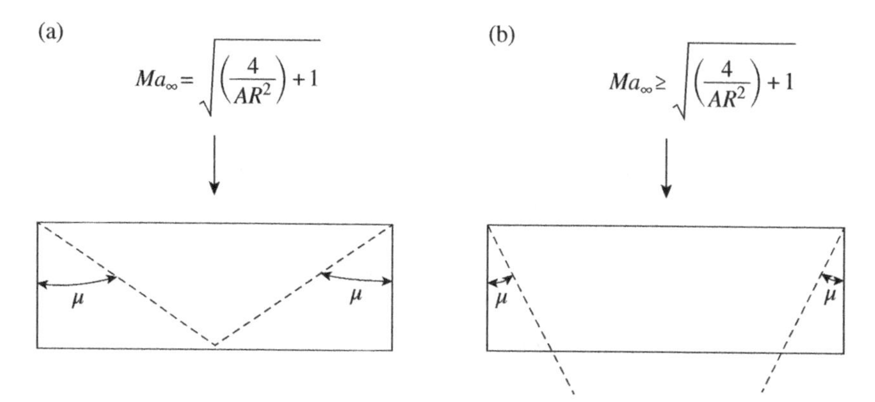


**Figure 9.4** Rectangular wing configuration and wing-tip Mach cones: (a) shows flow condition where three-dimensional regions meet at the center aft of the wing, and (b) shows three-dimensional regions for higher Mach numbers.

on the free stream Mach number ($Ma_\infty$) and on the aspect ratio (***AR***) of the wing. The Mach cone that issues from the forward wing-tip edges at each side only influences the wing region enclosed by the Mach angle, as seen in Figure 9.4. Inside these Mach cones the flow is three-dimensional – the rest of the span is not affected by the edges and operates under two-dimensional flow. Only at very low supersonic Mach numbers do rectangular wing planforms exhibit a series of overlapping regions that result from combined conical zones, and for many common wing configurations such cases appear only close to the transonic region. For example, in rectangular wings with $\boldsymbol{AR} > 3.0$ the three-dimensional cones from the wing's tips will not overlap when the incoming Mach numbers correspond to $\mu < 45°$. Figure 9.4 shows more general Mach number criteria that derive from purely geometrical considerations.

We proceed next with a special formulation for the lift in rectangular wings that accounts for both two- and three-dimensional regions. The smaller contributions from the three-dimensional regions can be simplified by assuming that inside each Mach cone at the wing tips the contribution to the lift is *one-half* the amount from the same area under two-dimensional flow conditions – this assumption is equivalent to *Busemann's second-order approximation theory for a flat-plate wing* (see Bertin and Cummings 2013). The result from this analysis, Eq. (9.11), is written in terms of our modified lift-slope factor for the total wing, which for wing sections was given the symbol $\xi_{\ell a}$. Our new dimensionless *total lift-slope factor* $\Xi_{\ell a}$ for finite wing flat-plate planforms is defined as:

$$\Xi_{\ell a} \equiv C_{La} Ma_\infty^2$$

Note the influence of the aspect ratio ***AR*** in Eq. (9.11) for conditions when conical regions at the edges do not overlap (i.e. that they do not exceed one-half of the span at the trailing edges). This criterion restricts the minimum free stream Mach number as indicated in the accompanying formula. Figure 9.5 is a plot of Eq. (9.11) for the indicated range of ***AR***s. The dash vertical line represents a transonic region boundary that needs to be omitted, so the plots begin at $Ma_\infty = \sqrt{2}$.

$$\Xi_{\ell a} \equiv \frac{d}{d\alpha}\left(\frac{2L}{\gamma p_\infty bc}\right) = \frac{2\boldsymbol{AR} Ma_\infty^2}{\sqrt{Ma_\infty^2 - 1}} - \frac{Ma_\infty^2}{Ma_\infty^2 - 1} \quad \text{and} \quad Ma_\infty \geq \sqrt{\left(\frac{4}{\boldsymbol{AR}^2}\right) + 1} \tag{9.11}$$

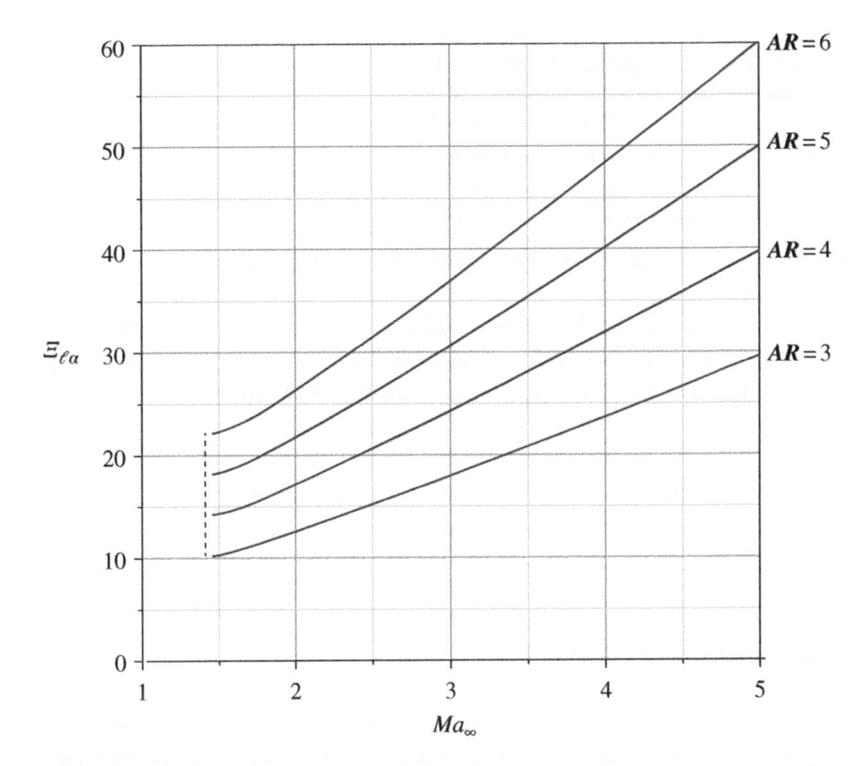


**Figure 9.5** Estimates of *supersonic three-dimensional lift-slope factor* $\Xi_{\ell\alpha} \equiv C_{L\alpha}Ma_\infty^2 = d(2L/\gamma p_\infty bc)/d\alpha$ for rectangular flat plates. Here conical-flow tip edges contribute half the lift amount of the two-dimensional flow region.

As shown in Figure 9.5, our three-dimensional lift-slope factor increases continuously with approaching Mach number and with rising values of the aspect ratio. This is in contrast to the Mach number dependence of the conventional two-dimensional dimensional lift coefficient itself, Eq. (9.8b), and illustrates how different non-dimensional formulations for the lift slope can suggest dissimilar messages.

**Example 9.5** Find the combination of atmospheric pressure and $\alpha$ that allows of an aircraft weighting $W_0 = 12\,000$ lbf to cruise at Mach numbers of 2, 3, and 4. Then find the compatible cruising altitude once an angle of attack is chosen. Use the rectangular wing performance in Figure 9.5 and take a wing planform of $b = 10.0$ ft and $c = 2.0$ ft.

The dimensionless coefficient as defined in Eq. (9.11) contains $2L/\gamma p_\infty bc$ and we know that $L = W_0$ and with the given wing coefficient $\boldsymbol{AR} = 5$. The following table shows results for the product $\alpha p_\infty = 2W_0/(\Xi_{\ell\alpha}\gamma bc) = 857/\Xi_{\ell\alpha}$. Once an angle of attack (in radians) is chosen, we can then find the altitude using an appropriate source. Here the last entry comes from a Standard Atmosphere table (heights shown are approximate) using $\alpha = 3° = 0.0523$ radians.

| $Ma$ | $\Xi_{\ell\alpha}$ | $\alpha p$ (lbf/ft$^2$) | $h$ (ft) for $\alpha$ = 0.0523 (3°) |
|---|---|---|---|
| 2.0 | 21.76 | 39.39 | 26 000 ($p_\infty$ = 753 lbf/ft$^2$) |
| 3.0 | 30.69 | 29.92 | 34 500 ($p_\infty$ = 534 lbf/ft$^2$) |
| 4.0 | 40.25 | 21.29 | 39 000 ($p_\infty$ = 407 lbf/ft$^2$) |

Higher angles of attack would have resulted in increases in the cruising altitude.

The *wave drag contribution* from a rectangular flat-plate wing at supersonic speeds follows suit. Equation (9.9) shows that the two-dimensional drag coefficient becomes $\alpha c_\ell$ so that for three-dimensional wing the following wave drag coefficient along with Eq. (9.11) applies:

$$C_{Dw} = \alpha C_L = \alpha^2 \Xi_{\ell\alpha}/Ma_\infty^2 \tag{9.12}$$

Several other contributions to the drag must be included when estimating the thrust needed from the propulsion unit as they are typically of the same magnitude as the wave drag.

## 9.6 Aircraft Wings in Compressible Flow

The Prandtl–Glauert rule can also be used with finite wings in subsonic flow, and the applicable pressure coefficient has the same form as Eq. (9.4). The coefficient for the total lift and associated pitching moment about the aerodynamic center will similarly be given by their incompressible formulas for nearly elliptic planforms from Chapter 6, Eqs. (6.17) and (6.21), both now over $\sqrt{1-Ma_\infty^2}$. Division by the same denominator applies to the induced drag, Eq. (6.13), and to the aspect ratio that becomes:

$$\boldsymbol{AR} = \frac{\boldsymbol{AR}_0}{\sqrt{1-Ma_\infty^2}} \tag{9.13}$$

### 9.6.1 Swept Wings

Modifying the wing's frontal geometry can remediate to the *drag divergence* experienced by aircraft as flow over their wings reaches the *critical Mach number* given by Eq. (8.25) and in Figure 8.7. *Swept wings*, seen in nearly all modern large aircraft, are designed to decrease this extra drag. In addition to skin drag and induced (or finite wing) drag, supersonic aircraft need to cope with contributions from wave drag and pressure drag that can be worsened by shocks that typically develop in the transonic region. In contrast to subsonic flow over an unswept wing that has exceeded the *critical Mach number*, the same flow over a swept wing sees a normal free stream vector component decreased by an amount cos($\Lambda$). The sweep angle of the leading edge ($\Lambda$) must be such that the flow perpendicular to the wing ($Ma_\infty$ cos($\Lambda$)) is below the critical Mach number value for that particular wing (see Figure 9.6). Aerodynamic effects are primarily governed by the flow normal to the wing's leading edge, although this is a simplification of a more complicated problem.

In unswept finite wings at supersonic speeds, only the wing tips and the center wing portion attached to the fuselage is subject to three-dimensional flow effects with the rest of the wing being under two-dimensional flow (see Section 9.5). Low aspect ratio (***AR***) thin wings designs can be efficient in supersonic flight since they reduce wave drag much more than increase the induced drag, whereas high aspect ratio wings are more efficient for subsonic flight as discussed in Chapter 6 for incompressible flows.

Supersonic aircraft also benefits from swept wings – when the leading edges of a swept wing are inside of the *Mach cone* of the free stream issuing from the aircraft's centerline (or wing spine), the component normal to the leading edge is subsonic and wave drag is not produced. For moderately *swept* flat-plate rectangular wings, Figure 9.5 may still be used by substituting $Ma_\infty \cos(\Lambda)$ for $Ma_\infty$ but there also needs to be account of a central three-dimensional region at the root-chord region (or wing spine) depicted in Figure 9.6 that becomes non-negligible in short spans. In order to avoid three-dimensional region overlaps with rectangular swept wings, the minimum Mach number needed for Eq. (9.11) to apply would now be given by:

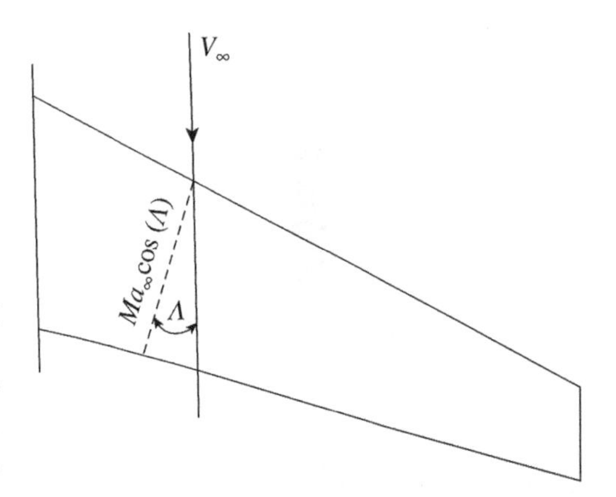


**Figure 9.6** Swept wing where $Ma_\infty \cos(\Lambda)$ is designed to be below the critical Mach number.

$$Ma_\infty \cos(\Lambda) \geq \sqrt{\left(\frac{4}{AR^2}\right) + 1}$$

Equivalent geometrical effects with ideal flows are obtained by sweeping the wing either backward or forward by the angle ($\Lambda$). In subsonic flows, the Prandtl–Glauert rule may also be applied to account for compressibility in the form (where "$\cot(\Lambda)_0$" is the incompressible value):

$$\cot(\Lambda) = \cot(\Lambda)_0 / \sqrt{1 - Ma_\infty^2} \tag{9.14}$$

The geometry for an aircraft that pivots its *entire wing* is shown in Figure 9.7. This *oblique wing aircraft* was conceived by R. T. Jones at NASA to explore managing wave drag with a single wing

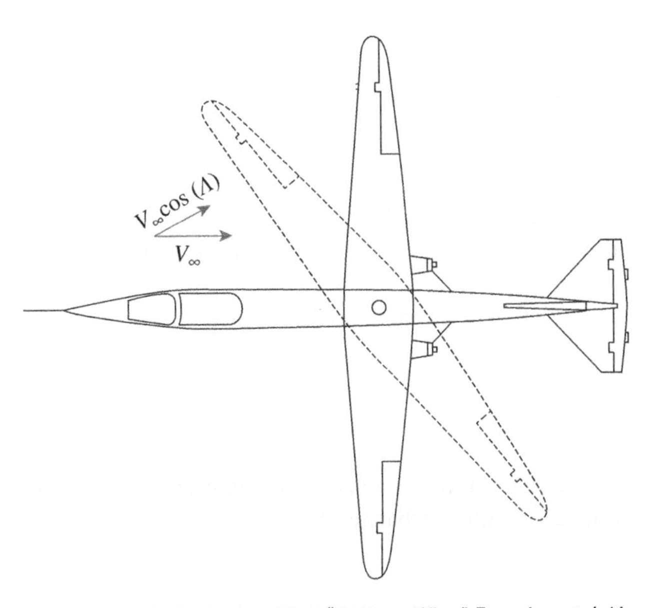


**Figure 9.7** NASA-Dryden AD-1 "Oblique Wing" Experimental Aircraft. *Source:* Adapted from The oblique wing aircraft design for transonic and low supersonic speeds, by R. T. Jones, Acta Astronautica 4 (1–2) (January–February 1977): 99–109.

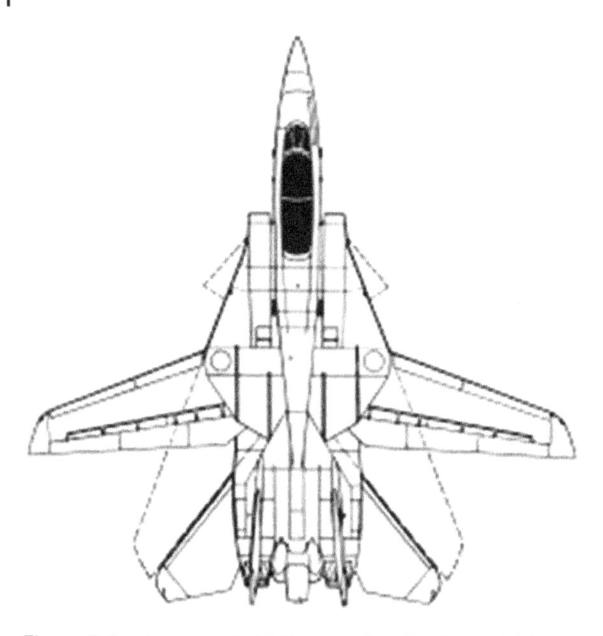

**Figure 9.8** Grumman F-14 Tomcat aircraft with variable position wings. The wings are spread out only during take-off and landing. [Search the web for Grumman F-14 Tomcat and for swept wing aircraft search for "swept wing aircraft images."]

that can rotate its flow orientation in a continuous manner thereby matching changes of Mach number. Figure 9.7 shows that the velocity component normal to the entire wing is completely free from the conventional center spine of symmetrically swept wings. The sweepback effect is forward on one side and backward on the other. This aircraft was successfully tested but never reached production. The oblique wing is able to match the planform area for lift = weight cruise requirements as Mach number increases without changes of angle of attack and affords easier vehicle parking with the wing folded for tight spot locations.

Several high-speed military aircraft have been designed with moveable wings to optimize performance for both their subsonic and supersonic Mach ranges. The F-14 in Figure 9.8 is a schematic showing how moveable wings can be symmetrically deployed; the wing forms a "delta configuration" when the front and rear wing surfaces are folded back to a sweepback angle of 68° at its cruising Mach number is 2.4. More modern military aircraft designs, however, such as the F-35 Joint Strike Fighter have no pivoting wings.

## 9.7 Enrichment Topic

### 9.7.1 Classical Forms of the Small Perturbation Equation

In this section we examine how the form of Eqs. (9.4) and (9.7) for the pressure coefficient may be obtained through manipulations of Eq. (9.1). This will hopefully help understand how Eq. (9.1) may be used to obtain other equations in this chapter. We consider two-dimensional flows in their purely subsonic or supersonic regimes and exclude the transonic region since it has mixed subsonic and supersonic regions.

With purely *subsonic flows* we can make an "affine coordinate transformation" that essentially stretches the $x$-dimension because $Ma_\infty < 1.0$.

$$\tilde{x} = \frac{x}{\sqrt{1 - Ma_\infty^2}} \tag{9.15}$$

This then results in the standard incompressible form of Laplace's equation after taking the $x$-derivative twice ($y$, the cross-flow coordinate, remains unchanged):

$$\frac{\partial^2 \phi}{\partial \tilde{x}^2} + \frac{\partial^2 \phi}{\partial y^2} = 0$$

Using the solution for the pressure coefficient in Eq. (9.3) leads to the form of the *Prandtl–Glauert rule*:

$$C_p = -\frac{2u}{V_\infty} = -\frac{2}{V_\infty}\frac{\partial\phi}{\partial\tilde{x}}\frac{d\tilde{x}}{dx} = \frac{C_{p0}}{\sqrt{1-Ma_\infty^2}}$$

As before, $C_{p0}$ is the incompressible value of the pressure coefficient. An extension to finite wings includes adding the term $\partial^2\emptyset/\partial z^2$ but neither $z$ nor $y$ need to undergo the transformation.

With *supersonic flows*, when non-dimensionalized along their *characteristic lines*, Eq. (9.1) leads to a comparably useful result. As stated earlier, the potential function $\phi$ has units of $m^2/s$ (or $ft^2/sec$) so that the product of $V_\infty$ with a properly chosen length constant renders $\phi$ dimensionless. Here we want to focus on the *natural* coordinates *of supersonic flow* – in Chapter 8 we introduced the *characteristic lines* or *Mach lines* that are intrinsic in supersonic flows; these depend on the free stream Mach angle ($\mu_\infty$) shown in Figure 8.1b as:

$$\tan^2\mu_\infty = \left(Ma_\infty^2 - 1\right)^{-1}$$

If we choose lengths $l_y$ and $l_x$ along characteristic lines for non-dimensionalizing then:

$$\tan^2\mu_\infty = \left(\frac{l_x}{l_y}\right)^{-2} = \left(Ma_\infty^2 - 1\right)^{-1}$$

In the small perturbation equation $l_x/l_y$ represents the slope of the characteristics:

$$\left(1-Ma_\infty^2\right)\frac{\partial^2\emptyset}{\partial\hat{x}^2} + \left(\frac{l_x}{l_y}\right)^2\frac{\partial^2\emptyset}{\partial\hat{y}^2} = \frac{\partial^2\emptyset}{\partial\hat{x}^2} - \frac{\partial^2\emptyset}{\partial\hat{y}^2} = 0$$

where $\hat{x} \equiv x/l_x$, $\hat{y} \equiv y/l_y$ and $\tan\mu_\infty = (l_y/l_x)$

Having absorbed the factor $\left(Ma_\infty^2 - 1\right)$ we can use generalized solutions like those of the conventional form of the wave equation. The following results from thin airfoil theory apply here (see Liepmann, 1957) for the two-dimensional pressure coefficient ($C_p$) at the airfoil:

$$C_{pu} = 2\left(\frac{d\hat{y}}{d\hat{x}}\right)_u = \frac{2}{\sqrt{Ma_\infty^2 - 1}}\left(\frac{dy}{dx}\right)_u \quad \text{Upper surface}$$

$$C_{pl} = -2\left(\frac{d\hat{y}}{d\hat{x}}\right)_l = \frac{2}{\sqrt{Ma_\infty^2 - 1}}\left(-\frac{dy}{dx}\right)_l \quad \text{Lower surface}$$

Approximations with thin airfoils make the local slope at the surface $dy/dx \equiv \tan\theta \approx \theta$, which for a flat plate becomes the angle of attack ($\alpha$) in Eq. (9.8a). Unlike subsonic flows, flow perturbations in supersonic flows have been shown to give rise to *wave drag*. Note that for supersonic flows some authors use another "$C_{p0}$" defined at $Ma_\infty = \sqrt{2}$, the Mach number where we found the lift curve to be a minimum according to Figure 9.2. This location is typically just beyond the transonic region and one that makes the supersonic form of Eq. (9.1) the traditional wave equation. By arguments similar to those already introduced, it can also be shown that in supersonic flows airfoil thickness only contributes to the drag.

## 9.8 Summary

We have discussed sectional and, for rectangular wings, total compressible flow functions for aerodynamic coefficients of interest. In three-dimensional supersonic flows, only the lift force was detailed because the drag has a number of contributions beyond wave drag that complicate the formulations. Thin flat plates have been featured because they are the best sectional configurations for supersonic flow since thickness and camber add to wave drag without contributing to the lift. Equation (9.1) is used to infer features that lead up the Prandtl–Glauert and Ackeret rules. By sweeping the wings in subsonic aircraft, drag divergence is delayed to higher cruise Mach numbers because swept wings can be constructed to operate below the critical Mach number and thus avoid supersonic shocks and their associated form drag increases.

The dimensionless coefficients defined in Chapter 2, while useful for incompressible flows, need to be modified for compressible flows. This is because the dynamic pressure $q_\infty$ found in their denominator contains the Mach number, and in this chapter we are interested in examining the behavior of these coefficients as a function of $Ma_\infty$. To properly represent compressible flows, we need to use the following coefficients:

$$\xi_\ell \equiv \frac{2\ell}{\gamma p_\infty c} = c_\ell Ma_\infty^2$$

$$\xi_m \equiv \frac{2m}{\gamma p_\infty c^2} = c_m Ma_\infty^2$$

$$\xi_{dw} \equiv \frac{2d_w}{\gamma p_\infty c} = c_{dw} Ma_\infty^2$$

The two-dimensional lift-slope factor:

$$\xi_{\ell\alpha} \equiv c_{\ell\alpha} Ma_\infty^2 = \mathrm{d}(2\ell/\gamma p_\infty c)/\mathrm{d}\alpha$$

introduced in Section 9.4 is more appropriate than $c_{\ell\alpha}$ itself for examining the dependence of the lift (and other parameters) in thin airfoils as a function of free stream Mach number. Note that both $c_\ell$ and $\xi_\ell$ are dimensionless, but that the latter contains no flow velocities. As seen in Figure 9.2, excluding the transonic region, $\xi_{\ell\alpha}$ increases continuously with $Ma_\infty$ eventually becoming linear for angles of attack below the stall condition.

The finite wing or three-dimensional lift-slope factor for a rectangular wing is:

$$\Xi_{\ell\alpha} \equiv C_{L\alpha} Ma_\infty^2 = \mathrm{d}(2L/\gamma p_\infty bc)/\mathrm{d}\alpha$$

It is also defined to be independent of the Mach number. Equation (9.11) and Figure 9.5 feature rectangular flat-plate planform behavior as function of $Ma_\infty$ and ***AR***. Results from our examples show that cruising at speeds where compressibility becomes manifest requires less planform area than incompressible speeds for equivalent weights, and this is one reason why several high-speed aircraft were designed with moveable wings.

## Problems

**9.1** Plot the slope of the lift coefficient ($c_{\ell\alpha}$) for both Prandtl–Glauert and Ackeret rules as in Figure 9.2. In Eq. (9.5b) take $c_{\ell 0} = 2\pi\alpha$. The values at $Ma_\infty = 1.0$ must be avoided in this plot. Compare your results to Figure 9.2.

**9.2** Using the Prandtl–Glauert rule calculate the lift coefficient ($c_\ell$) for a symmetric airfoil both at $\alpha = 6°$ and $Ma_\infty = 0.2$ and 0.7.

**9.3** Using the formulation for the pressure coefficient from Ackeret theory, Eq. (9.7)), calculate the pressure coefficient for the first set of waves at the isosceles triangle in Figure P9.3. Compare the answers to using oblique shock results for the pressure ratios.

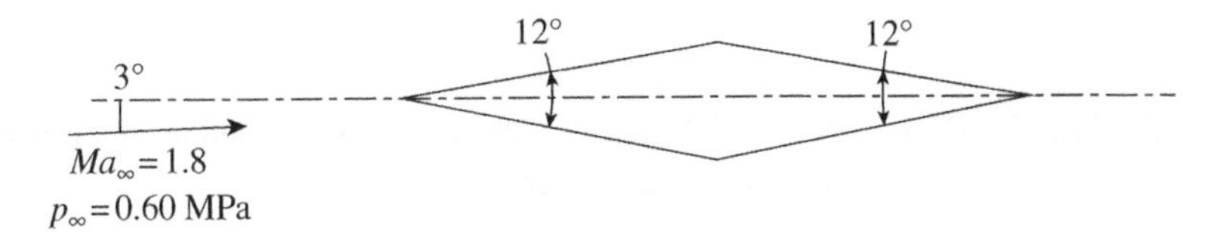


**Figure P9.3**

**9.4** Plot the *Mach angle* as a function of Mach number from $Ma_\infty = 1.0$ to 7.0. Identify the *transonic, supersonic,* and *hypersonic* regions. What Mach number approaching a rectangular wing of $\boldsymbol{AR} = 6$ would result on no more than 10% of the planform area ($S = bc$) under a three-dimensional flow situation?

**9.5** The wing in Example 9.5 is used in a supersonic aircraft that cruises at an altitude of 40 000 ft with $\alpha = 3°$. At this operating Mach number, it has been found that the wave drag from the wing amounts to about 40% of the total aircraft drag. Estimate the thrust required from the engines to maintain this flight condition.

**9.6** When an airfoil section is part of a wing with a 30° sweepback angle, the critical Mach number of the wing is found to be 0.8. Estimate the critical for another wing with the same section but with sweepback 43°.

**9.7** A rectangular wing operates at $Ma_\infty = 4.0$. The wing aspect ratio is 2.6.
a) Over what fraction of the wing does 2D flow exist? Draw a sketch.
b) Assuming 2D flow over the entire wing, give the value for the slope $C_{L\alpha} = c_{\ell\alpha}$.

**9.8** A Prandtl–Meyer expansion takes place from $Ma_1 = 2.13$, $p_1 = 10$ psia, with a turn angle of 4°. Considering the flow to turn over a sharp corner, calculate $C_p$ after the expansion by two methods, namely, the small perturbation theory and the Prandtl–Meyer theory. If a noticeable difference is apparent, which one is more correct?

**9.9** A cambered finite wing operating at $Ma_\infty = 0.45$ at an angle of attack of zero lift of −3° and $C_{L\alpha} = 5.3$ (per radian). What will be the lift coefficient for $\alpha = 5°$ and $Ma_\infty = 0.7$? The wing is elliptical, unswept with $\boldsymbol{AR} = 9$.

**9.10** A two-dimensional supersonic airfoil operates at $Ma_\infty = 2.8$ and $\alpha = 5°$. If the airfoil now decelerates to a new cruise condition $Ma_\infty = 2.3$, what should the new AOA be? Assume the following change in operating altitude from 10 200 to 8 400 m.

**9.11** A wing in subsonic flow operates at $Ma_\infty = 0.4$ and has an aspect ratio of 6.5. The two-dimensional, incompressible lift-curve slope is 90% theoretical. If the wing is unswept with a zero-lift angle of attack of −1.5°, what is the lift coefficient at $\alpha = 8°$?

**9.12** Show that $\emptyset = (\tilde{x})^2 - y^2$ where $\tilde{x} = x\left(1 - Ma_\infty^2\right)^{-1/2}$ satisfies Eq. (9.1) – the compressible form of Laplace's equation.

## Check Test

**9.1** Explain why the Prandtl–Glauert rule and Ackeret's rule cannot be applied in the transonic range using *both* physical and mathematical arguments.

**9.2** Explain why the best airfoil configuration in supersonic flow is a thin, flat-plate contrasting the reasons camber and thickness are only desirable in subsonic airfoils.

**9.3** Does the statement that two-dimensional rectangular airfoils are equivalent to those with infinite ***AR***s apply to both subsonic and supersonic airfoils?

**9.4** For a flat-plate airfoil, what are the lift and wave drag *coefficients* $c_\ell$ and $c_{dw}$ at $Ma_\infty = 2$ and $\alpha = 5°$? Compare the trend of these answers to those for the coefficient $\xi_\ell$ and for $\alpha\xi_\ell$ as the Mach number rises. What can you conclude is the actual trend of the lift and drag coefficients with increasing $Ma_\infty$?

**9.5** A theoretical, two-dimensional, flat-plate airfoil operates at $Ma_\infty = 0.5$ and $\alpha = 7°$. If we want to operate at the same altitude but at $Ma_\infty = 1.65$ what should the angle of attack be?

**9.6** A thin, two-dimensional symmetric airfoil is tested at $Ma_\infty = 0.58$ and found to produce $c_\ell = 1.4$ at $\alpha = 5°$. What lift coefficient would you expect to measure for this airfoil in a low speed wind tunnel at $\alpha = 7°$?

**9.7** Start with the definition of the pressure coefficient and show that $p = p_\infty \left[1 + \frac{C_p \gamma Ma_\infty^2}{2}\right]$.

**9.8** What is the pressure coefficient at an airfoil station where the local Mach number is 0.9 if $Ma_\infty = 0.6$?

**9.9** Compute $c_\ell$ and $c_d$ for a two-dimensional flat-plate airfoil at $\alpha = 5°$ for $Ma_\infty = 0.5$, 1.0, and 1.5. Neglect viscous effects.

**9.10** For subsonic flow over airfoils, name two conditions/limitations that would cause the Prandtl–Glauert rule to produce inaccurate results.

**9.11** Name two design changes that can be made to increase the critical Mach number of a finite wing.

# 10

# Transonic and Hypersonic Aerodynamics

## 10.1 Introduction

In this chapter we examine two separate flight regimes that while based on the same overall fluid dynamic principles are sufficiently different from other aerodynamics regimes to merit separate treatment. Transonic and hypersonic flows are also different from each other and need to be treated in a more advanced manner than our coverage of flow regimes in previous chapters. Their importance is nearly matched by their complexity, so we will only consider topics that can be presented without the aid of numerical tools. Our study of transonic and hypersonic flows will focus on some aspects of their unique character. Formulations with compressible-flow boundary layers and high temperature gas effects are beyond the scope of our presentation.

Supersonic flight is presently routinely achieved, but to get there the aircraft needs to pass through the transonic region (Mach 0.7–1.4), one that generates high drag forces and often significant flow unsteadiness thereby challenging thrust production from ordinary low-speed powerplants and requiring advanced controls on the aircraft. Recall that when going through Mach 1, the *aerodynamic center* location, or "neutral point" on the wing (where the moment coefficient is independent of angle of attack), changes from the quarter chord to the half chord. With transonic flows we need to depart from our default thin airfoils' assumptions because research has shown that for cruising at this regime "supercritical airfoils" need to be thick to minimize overall drag unlike more conventional wing designs. We will examine the supercritical airfoil at a discrete airfoil location, namely, the upper aft region where shocks and boundary layer separation most often occur.

The hypersonic regime (Mach 5 and beyond) is challenging because of an emergent importance of thermal phenomena that result from vigorous frictional heating inside the boundary layers and from much stronger shocks, both of which have had effects that could be neglected on the gas temperature up to now. One consequence of hypersonic heating is that the specific heat ratio ($\gamma$) does not remain constant throughout the flow since in air it depends on temperature. Not surprisingly, analyses of hypersonic flows are more demanding because the equations that describe them are highly nonlinear, and experimental facilities require relatively more expensive and sophisticated equipment. In hypersonic flows, highly empirical correlations and advanced software become indispensable tools.

## 10.2 Objectives

After completing this chapter, you should be able to:

1) State reasons why transonic and hypersonic flows are different from other flow regimes.

*Elements of Aerodynamics: A Concise Introduction to Physical Concepts*, First Edition. Oscar Biblarz.

Companion website: www.wiley.com/go/elementsofaerodynamics

2) Explain the "bridging role" of the nonlinear term in the transonic small perturbation form of Eq. (10.1) in terms of flow phenomena passing through $Ma = 1.0$ over a thin airfoil.
3) Justify why the *Prandtl–Meyer expansion* formulation given in Chapter 8 may be applied across transonic, supersonic, and hypersonic flow ranges.
4) Reason why outside the boundary layers normal *transonic shocks* may be treated as nearly isentropic.
5) Identify the main source increased drag in objects going through the transonic range.
6) Explain why through the transonic range the aerodynamic center must move from quarter-chord under subsonic conditions to half-chord under supersonic conditions.
7) Identify reasons for using thick or *supercritical* wings for transonic flow vehicles.
8) Identify what makes the ratio of specific heats for air depend on temperature.
9) Describe the reasoning behind the Newtonian hypersonic lift and drag formulas and how results compare to those from formulations developed in Chapters 8 and 9.

## 10.3 Transonic Flow

In Chapters 8 and 9 we examined the linear portion of the *small perturbation equation* introduced in Chapter 3 for thin airfoils in two-dimensional flows. The properties of linear *elliptic* and *hyperbolic partial differential equations* no longer apply in transonic flows because, as seen in Eq. (10.1), the transonic form conforms to neither type. Recall that elliptic behavior is synonymous with purely subsonic flow regions and hyperbolic behavior with purely supersonic domains, and note that the transonic region must somehow bridge these two.

$$\left(1 - Ma_\infty^2\right)\frac{\partial^2 \emptyset}{\partial x^2} + \frac{\partial^2 \emptyset}{\partial y^2} = \frac{Ma_\infty^2\,(\gamma + 1)}{V_\infty}\frac{\partial \emptyset}{\partial x}\frac{\partial^2 \emptyset}{\partial x^2} \tag{10.1}$$

Equation (10.1) is based on the assumption that all flow velocity perturbations are small compared to $V_\infty$ but, even so, as the approaching Mach number gets close to 1.0 the nature of the equation begins to change in significant ways. To better see this we rewrite (10.1) as follows:

$$\left[\left(1 - Ma_\infty^2\right) - \frac{Ma_\infty^2\,(\gamma + 1)}{V_\infty}\frac{\partial \emptyset}{\partial x}\right]\frac{\partial^2 \emptyset}{\partial x^2} + \frac{\partial^2 \emptyset}{\partial y^2} = 0$$

In purely subsonic or supersonic flows, the magnitude of $\left(1 - Ma_\infty^2\right)$ is much greater than the second term in the square bracket, but when $Ma_\infty$ ranges between 0.7 and 1.4 both terms inside the bracket become comparable as $\left(1 - Ma_\infty^2\right)$ shrinks in magnitude and flips sign. Depending on approaching Mach number and the shape of the perturbing object, the overall sign of the square bracket may transition between plus and minus, changing the nature of the equation between elliptic and hyperbolic at different airfoil locations. This is one reason why transonic flow has considerably richer behavior, regularly developing flow conditions that can lead to normal shocks, drag rise, and "aerodynamic buffeting." Analyses in this regime have been primarily numerical, though in Examples 10.1 and 10.2 we present some analytical features of a sonic solution. In Section 10.6 we elaborate further on the mixed mathematical behavior of Eq. (10.1).

Figures 8.8 and 10.1 show alternative depictions of the transonic domain. *Transonic shocks* are commonly treated as normal shocks that may be calculated from the relation:

$$(Ma_1 - 1) \approx (1 - Ma_2)$$

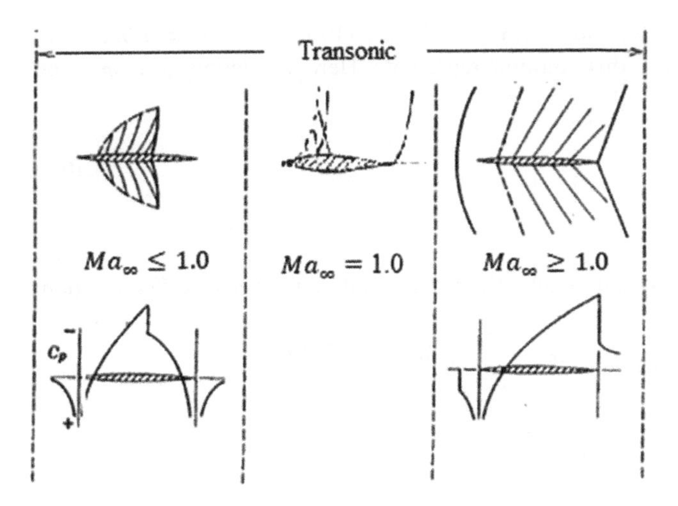


**Figure 10.1** Transonic flows crossing Mach one. *Source:* Adapted from Liepmann and Roshko (1957).

where the subscript 1 is the Mach number before the shock and 2 after it. This is readily verifiable by consulting a normal shock table for air up to $Ma_1 \leq 1.10$ (see Problem 10.1). A ratio that better reflects shock strength (or feebleness) under transonic conditions is the ratio of the entropy change across the shock divided by the ideal-gas constant ($\Delta S/R$) – a dimensionless entropy increase tabulated in some normal shock tables. Being very close to Mach 1.0, transonic shocks modify air properties (such as the velocity and stagnation pressure) in relatively limited ways; these shocks are, however, very effective at detaching the boundary layers at adverse pressure gradient regions. Such detachments, primarily observed on the upper rear portion of airfoils, are one of the main causes for steep form drag increases once termed "the sound barrier."

The compressible-flow rules of Prandtl–Glauert and Ackeret introduced in Chapter 9 cannot be applied in the transonic region because as already noted their associated sectional pressure coefficients become infinite at Mach 1. Experimental observations of the pressure coefficient show the existence of finite values in the transonic region. As given in Eq. (10.2), small perturbation theory predicts a fixed constant slope for the pressure coefficient in the sonic transition region. Further details on transonic flows are found in Guderley (1962) and Ramm (1990).

$$\left(\frac{dC_p}{dMa_\infty}\right)_{Ma_\infty = 1} = \frac{4}{\gamma + 1} \approx 1.67 \text{ for air} \tag{10.2}$$

This relation has been verified experimentally and gives additional support to the form for the velocity potential in Eq. (10.1). In Examples 10.1 and 10.2 we explore further intrinsic features of Eq. (10.1) by discussing useful results from an exact solution given in Biblarz (1976) and Biblarz and Priyono (1994).

Small perturbation theory is also useful for finding transonic ideal contours once a solution to Eq. (10.1) is known. To find a physical surface, we use Eq. (10.3) that stipulates a surface tangency with the flow. In other words, the streamlines must be parallel to their physical boundaries. In Example 10.1(b) we make use of this formulation.

$$\left(\frac{dy}{dx}\right)_{surface} = \frac{\varphi_y}{V_\infty} \tag{10.3}$$

**Example 10.1** (a) Show that Eq. (E10.1) for $\varphi(x, y)$ satisfies Eq. (10.1) when $Ma_\infty = 1.0$, and (b) find the physical surface profile that this solution represents. Here we denote $\emptyset$ as $\varphi$ when $Ma_\infty = 1.0$.

$$\varphi(x, y) = \frac{V_\infty}{3(\gamma + 1)} \frac{x^3}{y^2} \tag{E10.1}$$

**Solution**

a) Given Eq. (E10.1) for $\varphi(x, y)$, proof that it satisfies the sonic condition is given by differentiation:

$$\text{Show } \varphi(x, y) = \frac{V_\infty}{3(\gamma + 1)} \frac{x^3}{y^2} \quad \text{is a solution of} \quad \frac{\partial^2 \emptyset}{\partial y^2} = \frac{(\gamma + 1)}{V_\infty} \frac{\partial \emptyset}{\partial x} \frac{\partial^2 \emptyset}{\partial x^2}$$

$$\frac{\partial \varphi}{\partial x} = \frac{V_\infty}{(\gamma + 1)} \frac{x^2}{y^2} \quad \text{and} \quad \frac{\partial \varphi}{\partial y} = -\frac{2}{3} \frac{V_\infty}{(\gamma + 1)} \frac{x^3}{y^3}$$

$$\frac{\partial^2 \varphi}{\partial x^2} = \frac{2V_\infty}{(\gamma + 1)} \frac{x}{y^2} \quad \text{and} \quad \frac{\partial^2 \varphi}{\partial y^2} = \frac{2V_\infty}{(\gamma + 1)} \frac{x^3}{y^4}$$

$$\frac{(\gamma + 1)}{V_\infty} \frac{\partial \varphi}{\partial x} \frac{\partial^2 \varphi}{\partial x^2} = \frac{(\gamma + 1)}{V_\infty} \left[ \frac{V_\infty}{(\gamma + 1)} \frac{x^2}{y^2} \frac{2V_\infty}{(\gamma + 1)} \frac{x}{y^2} \right] = \frac{2V_\infty}{(\gamma + 1)} \frac{x^3}{y^4}$$

$$\frac{\partial^2 \varphi}{\partial y^2} = \frac{(\gamma + 1)}{V_\infty} \frac{\partial \varphi}{\partial x} \frac{\partial^2 \varphi}{\partial x^2}$$

which proves that with Eq. (E10.1) both sides of Eq. (10.1) become equal when $\left(1 - Ma_\infty^2\right) = 0$.

b) To find a physical surface, we need to use Eq. (10.3) that applies the surface tangency (i.e. it requires that the slope at the physical surface be the same slope of an adjacent streamline). Letting $x = 0$ when $y = y_0$,

$$\left(\frac{dy}{dx}\right)_{surface} = \frac{\varphi_y}{V_\infty} = -\frac{2}{3(\gamma + 1)} \frac{x^3}{y^3}$$

$$\int_{y_0}^{y} y^3 dy = -\frac{2}{3(\gamma + 1)} \int_0^x x^3 dx$$

$$\frac{y(x)}{y_0} = \sqrt[4]{1 - \frac{2x^4}{3(\gamma + 1) y_0^4}} \quad \text{where} \quad \frac{x}{y_0} \le \sqrt[4]{\frac{3(\gamma - 1)}{2}} \le 0.88$$

The preceding last entry identifies the applicable range of $x/y_0$ as $\le 0.88$ since $\gamma = 1.4$ for air. Figure E10.1 shows the resulting surface. This profile may represent part of an "end section" or "aft-contour" on the upper surface of a *supercritical airfoil* when a locally sonic flow (i.e. $Ma$ = 1.0) approaches this adverse pressure gradient region. In Section 10.4, Whitcomb's supercritical airfoil is pictured as a thick airfoil example, and a portion of the profile in Figure E10.1 is shown as "patched" on the upper rear section (refer to Figure 10.2).

With Example 10.2 we proceed to show that Eq. (E10.1) of Example 10.1 does represent a transonic expansion from $Ma = 1.0$, that is, it contains a Prandtl–Meyer wave dependance. Retaining the first two terms of a series expansion for the $\tan^{-1}(...)$ in Eq. (8.12) (i.e. near the sonic point

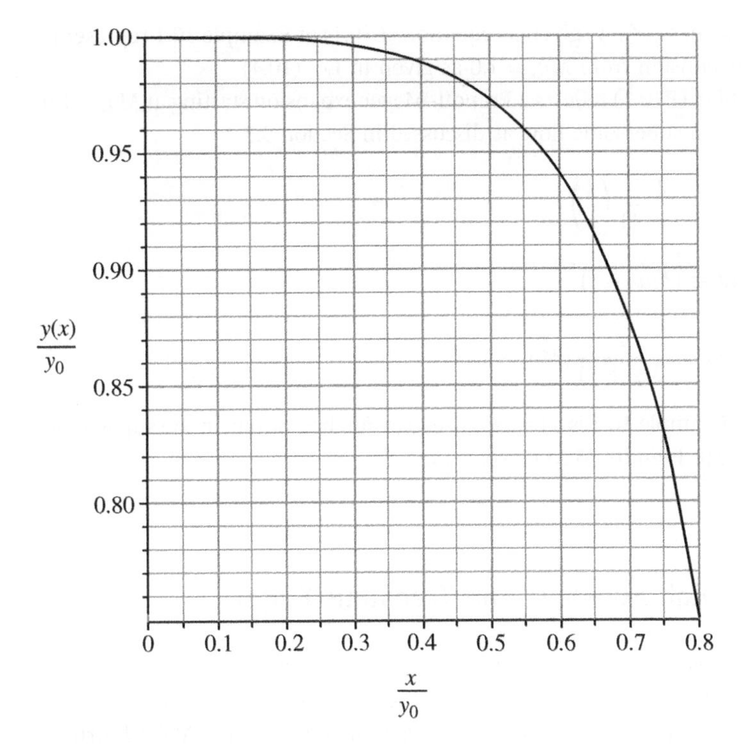


**Figure E10.1** Upper aft-contour sectional profile for the $Ma_\infty = 1.0$ solution to the transonic equation (E10.1) in Example 10.1(b).

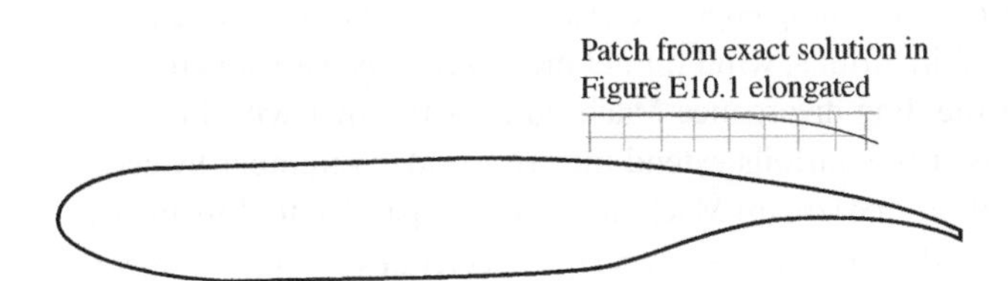


**Figure 10.2** Supercritical wing profile. The graph insert shown at the top right has a different aspect ratio than Figure E10.1 so as to compare it with the wing shown below it.

onset), the Prandtl–Meyer angle expression defined in Chapter 8 may be written as Eq. (10.4) (from Liepmann and Roshko 1957).

$$\nu \approx \frac{2}{3}\frac{(Ma^2-1)^{3/2}}{(\gamma+1)} + \ldots \tag{10.4}$$

This equation is consistent with this angle being zero at the sonic point. We need another relation from *small perturbation theory* (Liepmann and Roshko 1957), namely,

$$\tan(\nu) = -\frac{\varphi_y}{V_\infty}$$

to show the correspondence with our solution of $\varphi(x, y)$ from Example 10.1.

**Example 10.2** Show that the form of $\varphi(x, y)$ given as equation (E10.1) of Example 10.1 represents the onset of a Prandtl–Meyer expansion from $Ma_\infty = 1.0$ as given in Eq. (10.4).

In the following we show that Eq. (E10.1) reflects a Prandtl–Meyer expansion starting at $Ma_\infty = 1.0$ by examining characteristic lines in supersonic flow as discussed in Section 9.7.

$$\tan \nu \approx \nu = -\frac{\partial\varphi/\partial y}{V_\infty} = -\frac{2}{3(\gamma+1)}\left(\frac{x}{y}\right)^3$$

$$\left(\frac{x}{y}\right)_{characteristics} = \text{constant} \equiv \left(Ma^2 - 1\right)^{1/2}$$

$$\text{Along the characteristics: } \nu = \frac{2}{3(\gamma+1)}\left(Ma^2 - 1\right)^{3/2}$$

which confirms that the flow in Example 10.1 is a transonic expansion beginning on the supersonic side of the sonic point, i.e., a weak Prandtl–Meyer expansion.

## 10.4 Thick Airfoils for High Subsonic and Transonic Flight

Up to now in this book we have considered only *thin airfoils* because their properties are well understood. Here we depart and discuss a *thick airfoil* that has become more practical for aircraft in the high subsonic range. The *supercritical airfoil* and its general aviation version, the GA(W)-1 airfoil, have a relatively high maximum-thickness-to-chord ratio ($t_m/c$) designed to reduce the sharp drag increase that airplanes experience around Mach 1 – an increase primarily fed by shock-boundary-layer separation. Supercritical airfoils have a large well-rounded leading edge followed by flat top surface and bottom contours with a drooped trailing edge necessary to produce a sharp end point, see Figure 10.2. Compared to standard thin airfoils, wind tunnel measurements with supercritical airfoils indicate about a 15% delay in the drag divergence Mach number together with improvements of $C_{Lmax}$ at high subsonic speeds. These airfoils extend the critical Mach number by accelerating the flow to supersonic conditions at free stream Mach numbers comparable to thin airfoils, but when flown around their design condition their unique shape decelerates the upper flow to subsonic values through a series of very weak compression waves instead of the stronger shocks that form in conventional wing profiles.

Much of the supercritical airfoil development stemmed from analytical and experimental work is attributed to R.T. Whitcomb (who originated the *area rule* for transonic aircraft bodies – the implementation of this rule reduces flow losses at the junction of swept wings with the aircraft's fuselage) (see Anderson 2017; Bertin and Cummings 2013; Shevell 1983). To extend the drag-divergence Mach number, the *critical Mach number* of the airfoil has to be delayed and this has been done by appropriately modifying its sectional configuration. With the *supercritical airfoil* depicted in Figure 10.2, at its fairly blunt forward end the flow quickly accelerates on both top and bottom to low transonic condition. As the flow proceeds on the top of the airfoil, a supersonic region begins to form near the crest that remains weak by the flatness of the surface. Any shocks that develop on the top remain relatively tenuous so as not to detach the boundary layers thus delaying the "sound barrier" to higher approaching Mach numbers. Because the upper portion is much fatter than the bottom, the forward 2/3 portion of the airfoil represents a negative camber line that acts to decrease the lift at any given angle of attack. To compensate for this, the lower surface of the airfoil has a steep positive camber ending in the form of the cusp at its rear 1/3 as shown. Figure 10.2 also contains a "stretched-out version" of Figure E10.1 to visually compare our "patched shock-free segment" from the solution in Example 10.1.

For aircraft that cruise at high subsonic conditions, swept wings have been implemented that are appropriately thick. And for aircraft that fly supersonically, a variety of variable-swept thin-wing designs are available as shown in Chapter 9, Figures 9.6–9.8. All the these factors have meaningful implications for aircraft maneuvers and for propulsion system requirements during transonic flight.

## 10.5 Hypersonic Flow

Unlike supersonic flows, the hypersonic flow regime has no sharp lower bound so they are commonly referred as flows with $Ma_\infty > 5$. This domain covers Mach numbers where several rather complicated fluid-dynamic phenomena start to become important. For hypersonic airfoils flat plates remain a desirable sectional wing configuration, but for aircraft there are many advantages in designs where the entire body of the vehicle generates lift; this is because under hypersonic conditions wings, fuselage, and engines must be highly integrated to address prevailing heating and propulsion hurdles. In this regime it is necessary to evaluate the location and strength of shocks more carefully because, while they may be calculated with the incoming free stream value of the ratio of specific heats ($\gamma$), flow regions that follow hypersonic shocks exhibit significantly higher temperatures in addition to high pressure changes.

Over hypersonic wing sections Mach waves and oblique shocks tend to develop much closer to the surface forming what has been termed a "hypersonic boundary layer" where flow disturbances that surround the airfoil remain confined in direct analogy to viscous boundary layers. The flow outside such hypersonic boundary layers continues relatively unaffected by compression and expansion waves issuing from the immersed object. Moreover, since pressure increases from oblique shocks that form below an airfoil at angle of attack are very large relative to expansions that form above it, net pressure differences across the airfoil tend to reflect mostly lower surface values. All these observations are prelude to the applicability of a hypersonic analysis model based on Isaac Newton's work that is discussed later in this chapter.

At Mach numbers greater than 5, the Mach angle $\mu$ begins to be of the same order of magnitude as the flat-plate deflection angle $\alpha$ of a lifting body. Since the Mach angle is given by:

$$\mu \approx \sin(\mu) = 1/Ma_\infty$$

We may deduce that for a flat plate at angle of attack under hypersonic conditions:

$$Ma_\infty \alpha \geq 1.0$$

In most of the references neither oblique shocks nor Prandtl–Meyer expansions are properly tabulated in the hypersonic regime even though, as we have stated, the latter may be accurately calculated from Eq. (8.12) since expansion turning is isentropic and $\gamma$ remains constant. Downstream of hypersonic oblique shocks, however, gas temperatures, densities, and pressures calculated with Eqs. (8.7), (8.8), and (8.9) will need to reflect changes $\gamma$ that arise from gas heating.

### 10.5.1 High Mach Number Gas Effects

A schematic of waves and pressure coefficients typical to hypersonic flows is shown in Figure 10.3. This depiction is similar to the supersonic portion shown in Figure 9.1 except for the much shallower wave forms. Molecular components of air are mostly nitrogen and oxygen, both diatomic

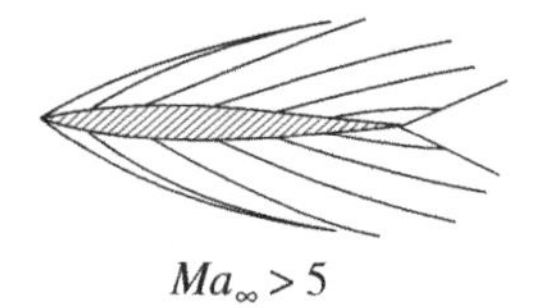


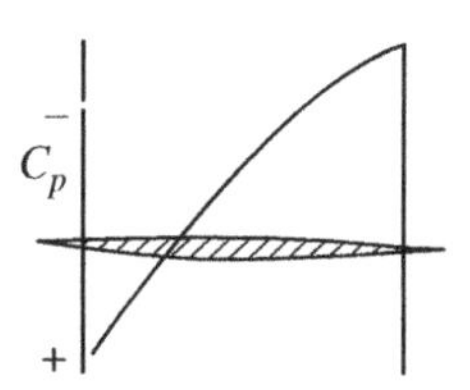


**Figure 10.3** Hypersonic flow around a thin airfoil. *Source:* Adapted from Liepmann and Roshko (1957).

gases. At atmospheric conditions the value of specific heats ($\gamma$) drops smoothly from 1.40 to 1.28 at temperatures of about 3550 K (or 6400 °R) but below thermal dissociation. Such a change results from the thermal activation of the vibrational mode of diatomic molecules. During *hypersonic re-entry* conditions, vehicles undergo such extreme heating that low atmospheric pressures lead to significant gas dissociation and chemical reactions in and around their exposed blunt regions. The space shuttle (presently retired) lifted off like a rocket and returned like a *blunt-body glider*. Bottom tiles were added to "thermally manage" and help the shuttle survive the severe heating rates generated during hypersonic re-entry. Vehicles designed to cruise at hypersonic conditions look different from those at lower flight regimes, having thin and needle-like front ends devoid of blunt front-facing regions. Both pointed- and wedge-nose designs operating as *lifting bodies* have been proposed – such types are called a *waveriders* (see Bertin and Cummings 2013; Anderson 2017) because they "surf upon" the shock waves that issue underneath of the lifting fuselage at hypersonic conditions. Propulsion for hypersonic air vehicles may be based on *rocket technology* (non-air breathing) for short ranges and on *scramjet technology* (air breathing but without rotating machinery) for longer vehicle ranges.

From a thermal perspective, other than stagnation regions, oblique shocks are a major source of heating. Frictional heating effects can also not be neglected because they affect all exposed surfaces. The supersonic flow-turning equations of Chapter 8 actually simplify somewhat at Mach numbers greater than 5 as discussed in this chapter.

### 10.5.2 Hypersonic Expansions

Prandtl–Meyer expansions required for convex turns actually cool the flow thus not affecting $\gamma$ (as long as air remains gaseous and does not condense). The Prandtl–Meyer function used for hypersonic flow calculations must reflect the $\gamma$ that applies to the incoming flow. (For hypersonic conditions the use of the *Aerodynamics Calculator* is recommended since most standard tables are not accurate in this range and do not give values other than $\gamma = 1.4$.)

**Example 10.3** Experimental observations of the lower aft portion of a flat-plate airfoil under hypersonic flow indicate that the turning angle is $\Delta\nu = 7°$ at the lower surface Prandtl–Meyer expansion, i.e. locations labeled regions 3–5 in Figure 8.9a. The air flow in region 3 where $Ma_3 = 5.28$ has been compressed through a leading-edge oblique shock so that its properties are $T_3 = 691.14$ K and $p_3 = 1.62\text{x}10^5$ N/m$^2$. Calculate gas properties after the Prandtl–Meyer expansion into region 5. Assume $\gamma = 1.4$.

At $Ma_3 = 5.28$ the ratios $p_3/p_t$ and $T_3/T_t$ together with $\nu_3$ are given as shown below. The Prandtl–Meyer calculations follow:

$$\nu_3 = 79.4° \quad p_3/p_t = 0.00137 \quad T_3/T_t = 0.1521$$

$$\nu_5 = 79.4 + 7 = 86.4° \quad \text{so that} \quad Ma_5 = 6.22 \quad p_5/p_t = 0.00051 \quad T_5/T_t = 0.114$$

$$p_5 = 0.00051 \times 1.62\text{x}10^5/0.00137 = 0.603\ \text{bar} \quad T_5 = 0.114 \times 691.14/0.152 = 518.4\ \text{K}$$

These results for the *second isentropic expansion at the trailing edge* show that at $T_5$ we may still retain $\gamma = 1.4$. Note that here gas properties differ from those of the free stream because regions 2 through 5 are isolated from the surrounding atmosphere by an envelope of various kinds of waves (not shown in Figure 8.9a).

### 10.5.3 Hypersonic Compressions

Hypersonic compressions resulting from oblique shocks can generate considerable heating downstream of the shock front but, since gas temperatures lag this gas heating at such high flow speeds, $\gamma$ remains basically unchanged *across* the shock front itself. Moreover, it can be seen from the isentropic formulas in Chapter 8 that the pressure is less sensitive than either the temperature or the density to small changes of $\gamma$ for any given Mach number. Equations (8.21) and (8.22) may be used to calculate effects of hypersonic oblique shocks provided that the final air temperatures stay below 3550 K (or 6400 °R) since $\gamma$ cannot change appreciably after the shock. Hypersonic flight normally takes place in upper layers of the atmosphere where temperatures are relatively low to begin with, and these temperature limits are seldom exceeded (except for blunt bodies during hypersonic re-entries).

For oblique shock waves under hypersonic conditions where free stream or incoming Mach number are generally above 5, Eq. (8.20) simplifies to become Eq. (10.5). Here $\theta$ is the oblique shock angle for weak shocks and $\delta$ the physical compression deflection angle (or $\alpha$).

$$\frac{\theta}{\delta} \approx \frac{\gamma+1}{4} + \sqrt{\left(\frac{\gamma+1}{4}\right)^2 + \frac{1}{(Ma_1 \sin(\delta))^2}} \tag{10.5}$$

**Example 10.4** Compare the $\theta$s calculated from Eq. (10.5) to those from Eq. (8.20) with $\gamma = 1.4$ at incoming Mach numbers of $Ma_1 = 10.0$ and 20.0 for $\delta = 10°$, 20°, and 30°.

| | $\theta$ at $Ma_1$ = 10.0 | | $\theta$ at $Ma_1$ = 20.0 | |
|---|---|---|---|---|
| $\delta$ | Equation (8.20) | Equation (10.5) | Equation (8.20) | Equation (10.5) |
| 10° | 14.43° | 14.31° | 12.71° | 12.66° |
| 20° | 25.82° | 25.34° | 24.70° | 24.35° |
| 30° | 38.51° | 36.97° | 37.54° | 36.25° |

As can be seen, the comparisons are quite close except at $\delta = 30°$ where the curves are becoming less linear. Equation (10.5) is therefore a relatively quick way to find $\theta$ in weak oblique shocks under hypersonic conditions.

In air flows below dissociation or chemical reactions, the dependance of the pressure ratio ($p_2/p_1$) across a shock to gas heating is seen in Figure 10.4 in the gap between the two surfaces. With such small $\gamma$-variations as shown, air temperature changes do not alter the pressure ratios significantly compared to Mach number changes or deflection angle changes. It should also be reiterated that at these high Mach numbers flow *non-equilibrium effects* become quite noticeable, a situation called "frozen composition flows" at exit flows from rocket nozzles where gas compositions significantly lag thermal flow changes that would exist in flows at *local equilibrium*. Being able to assume such frozen non-equilibrium greatly simplifies calculations but tends to underestimate real performance (see Sutton and Biblarz 2017).

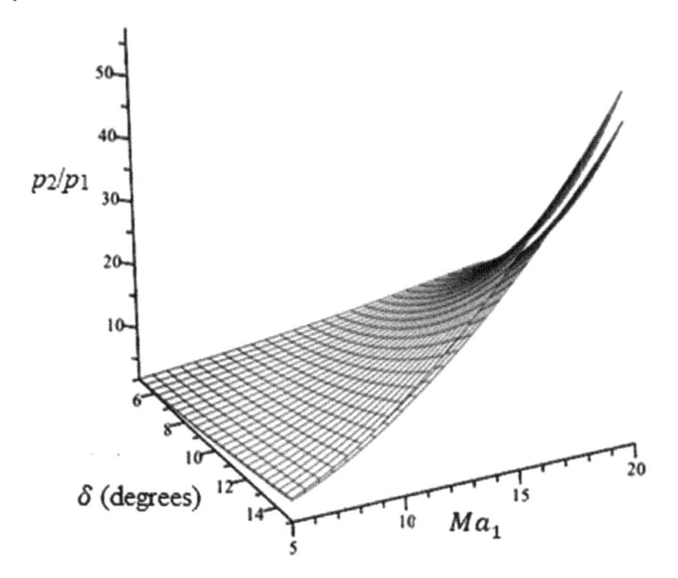


Figure 10.4 Pressure ratio ($p_2/p_1$) across weak oblique shocks in the hypersonic range as a function of incoming Mach number ($Ma_1$) and deflection angle $\delta$. The upper set of curves is for $\gamma$ = 1.4 and the lower set for $\gamma$ = 1.3.

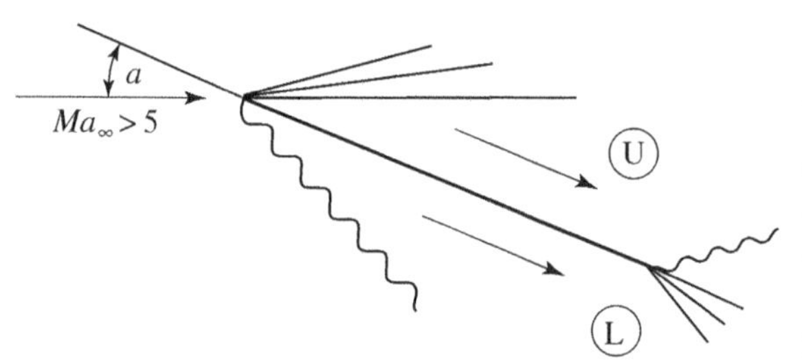


Figure 10.5 Thin flat plate under hypersonic flow conditions.

### 10.5.4 Hypersonic Flow Analysis "a la" Newton

At very high Mach number the pressure forces that develop over a flat plate appears to conform well with Newton's Corpuscular Theory that states that "the force of impact between a uniform stream of particles and a surface is obtained from the loss of momentum of the particles normal to the surface...". While not one of Newton's more famous laws, this formulation turns out to be a simple but practical method for analyzing forces in the hypersonic regime, one which increases in accuracy at Mach numbers above 10.

Applying Corpuscular Theory to a flat plate under hypersonic flow we designate the expansion at the upper surface with the subscript "U" (a location where the pressures become negligibly small) and to the oblique shock at the lower surface the subscript "L," see Figure 10.5. Since all waves develop very close to the flat plate, upon arriving at the inclined plate's surface the fluid's *normal component* of momentum is expected to vanish, whereas the *tangential component* will be preserved in direct analogy to the flow of a stream of tiny particles hitting a wall. From here the analysis proceeds with calculations of the resulting normal force on the plate. After some manipulation to represent the geometry of the inclined plate, we arrive at the following pressure coefficient description in Eq. (10.6):

$$C_{pU} \approx 0 \quad \text{and} \quad C_{pL} \approx 2\sin^2\alpha \tag{10.6}$$

where $\alpha$ here is the flat-plate angle of attack. The two-dimensional *force coefficient* normal to the plate ($c_n$) and its related lift and drag coefficients, all approximately, become Eqs. (10.7), (10.8), and (10.9).

$$c_n \approx 2\sin^2\alpha \tag{10.7}$$

$$\text{so that} \quad c_\ell \approx 2\sin^2\alpha\cos\alpha \tag{10.8}$$

$$\text{and} \quad c_d \approx 2\sin^3\alpha \tag{10.9}$$

In these formulas, the angle of attack of the flat plate is the only parameter that can vary and that makes Eqs. (10.8) and (10.9) particularly simple to use. It can also be further shown that the lift-to-drag ratio is the ratio of Eqs. (10.8)–(10.9) or

$$\ell/d = \cot(\alpha)$$

In Example 10.5 these formulations are compared against the more exact formulations we arrived at in Chapter 8.

**Example 10.5** Explore the resulting lift coefficients on a flat plate by comparing the oblique shock contribution using Eq. (8.21) to the Newtonian calculation using Eq. (10.8) in the range $Ma_\infty = 5$–$40$ to ascertain how constant the former becomes in the hypersonic range. The calculations of the lift coefficients are to be explored on a flat plate at two fixed angles of attack: (a) $\alpha = 15°$ and (b) $\alpha = 30°$. Explore also the effect of $\gamma$ changing from 1.4 to 1.3. Apply the notation of Figure 8.9a.

We will neglect any Prandtl–Meyer expansion contribution from the upper plate surface because it only slightly affects the low range of Mach numbers examined in this example. Under hypersonic conditions, flow expansion calculations yield vanishingly small values of the pressure $p_U$ (or $p_2$) at the upper flat-plate surface compared to the magnitude of $p_L$ (or $p_3$) after the oblique shock [check this for yourself]. Moreover, by neglecting the top pressures our results become more consistent with the assumptions in the Newtonian corpuscular model.

Using appropriate materials from Chapter 8 and Figure 8.9a for the oblique shock, the normal force coefficient may be written as:

$$c_n = \frac{2\left(\frac{p_3}{p_1} - \frac{p_2}{p_1}\right)}{\gamma Ma_1^2} \approx \frac{2\left(\frac{p_3}{p_1}\right)}{\gamma Ma_1^2}$$

$$c_\ell \approx \frac{2\left(\frac{p_3}{p_1}\right)\cos(\alpha)}{\gamma Ma_1^2}$$

Newtonian solution: at 15°, $c_\ell = 2\sin^2(15°)\cos(15°) = 0.129$ and at 30° $c_\ell = 2\sin^2(30°)\cos(30°) = 0.433$

Resulting oblique shock calculations using Equations 8.20 and 8.21 along with Newtonian values are shown in Figure E10.5 (a) and (b) where it can be seen that the oblique shock contribution flattens out beyond $Ma_\infty = 15$ but at values higher than the results from the Newtonian calculations.

Changes of $\gamma$ from 1.4 to 1.3 result in small deviations to the calculated pressures. There is about a 20% discrepancy with the Newtonian estimates in the flat portions of the upper and lower oblique shock curves, for both $\alpha = 15°$ and at $\alpha = 30°$; this suggests that this difference can be lessened by modifying the Newtonian constant. Note that Newtonian calculations are independent of $Ma_\infty$ and $\gamma$. Also, at Mach numbers below 10 the Prandtl–Meyer contribution on the airfoil's top surface is slightly more noticeable and this would make for somewhat better agreement with the Newtonian estimates.

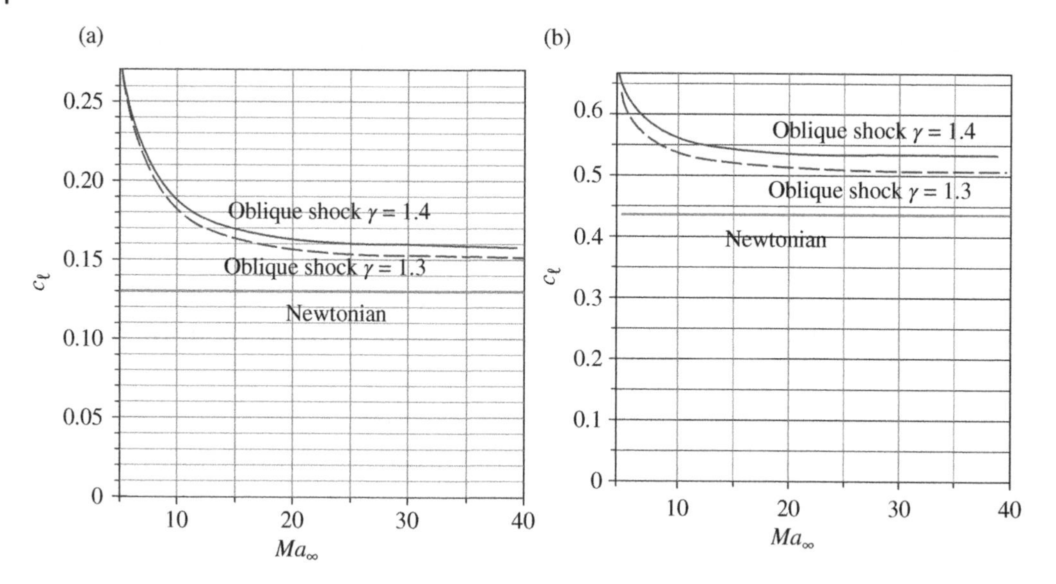


**Figure E10.5** The top solid curves are the oblique shock contribution to the flat-plate lift coefficient $c_\ell$ for regular air ($\gamma = 1.4$) and dashed curves below are for hot air ($\gamma = 1.3$), see also Figure 10.4. The lower straight lines are the Newtonian results. For (a) $\alpha = 15°$ and for (b) $\alpha = 30°$.

Flat-plate Newtonian calculations in Example 10.5 beyond $Ma_\infty = 15$ show the anticipated constant trend, but one whose value can be improved by replacing the factor "2" in Eqs. (10.6)–(10.9) with "2.4" and that would give in the preceding example $c_\ell = 0.155$ for $\alpha = 15°$ and 0.52 for $\alpha = 30°$, results very close to the oblique shock asymptotes in Figure E10.5a and b. Indeed, in Eq. (10.6) this constant is routinely modified to conform with experimental $C_p$-distributions around blunt bodies in hypersonic flows.

## 10.6 Enrichment Topics

### 10.6.1 Typifying the Transonic Equation

In mathematics courses we have been familiarized with Laplace's equation and the wave equation by designating or *typifying* such linear partial differential equations by their similar algebraic appearance to circles (called "degenerate or normalized ellipses") or hyperbolas. These similarities help visualize the harmonic nature of the former and the wave/characteristics nature of the latter. The transonic equation, however, is not a true "conic section" and attempts to represent the transition from elliptic to hyperbolic are harder to visualize.

In Figure 10.6 we have a multiple-curve figure depicting a degenerate ellipse (the dashed line), a hyperbola (the dash-dot line), and, enfolding some portion of each of these, a solid curve having an additional term made to represent Eq. (10.1). The solid curve is generated as an upside-down "cubic parabola" with the closed lower-half straddling the circle (or the subsonic domain) and the open upper-half straddling the hyperbola (or the supersonic domain). Note that in going from subsonic to supersonic this solid curve transitions between these two regions without discontinuities. As already stated, the left-hand portion of Eq. (10.1) may become either an ellipse or a hyperbola

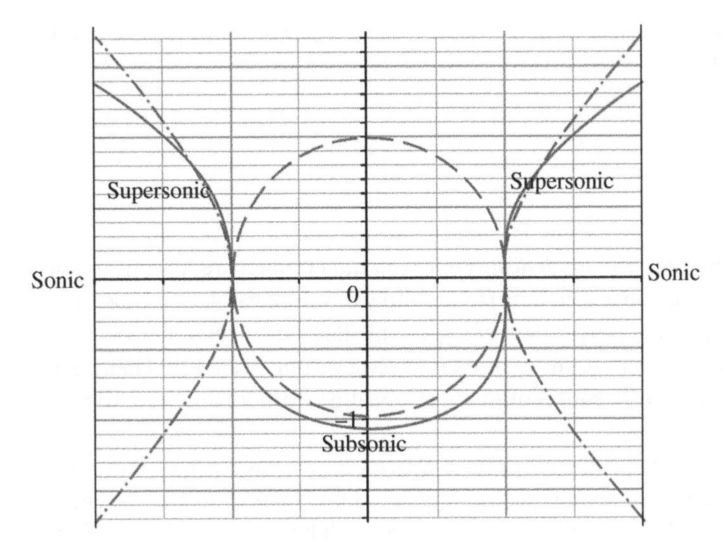


**Figure 10.6** Single curve depiction (closed below the sonic line and open above) that can straddle a *normalized* ellipse and a hyperbola. The dashed line is a closed curve and the dash-doted curves are open. Segments of the transonic equation represented by the solid curve are nearly elliptic where the factor $(1 - Ma_{\infty}^2)$ is positive at the bottom as the approaching flow is subsonic and nearly hyperbolic at the top where $(1 - Ma_{\infty}^2)$ is negative and the approaching flow is supersonic.

depending on the sign of $(1 - Ma_{\infty}^2)$ and the right-hand side nonlinear term facilitates the smooth transition between these two types of equation.

### 10.6.2 Effects of Elevating the Airflow Temperature Surrounding a Flying Vehicle

When the gas flow temperature locally increases above ambient conditions, the Mach number can be shown to decrease as $T^{-1/2}$ and the Reynolds number to decrease as $T^{-3/2}$. Higher temperatures occur intrinsically with hypersonic flows but can also be intentionally produced with laser-directed heating and similar properly focused external high energy inputs. The question becomes, are there any advantages that result from localized higher-temperature airflow sectors? We will restrict our comments to heating effects below air dissociation and ionization because these require considerably larger amounts of energy and special structural materials. Historically, however, interest in this subject originated under the heading *Plasma Aerodynamics*.

Locally focused heated regions are likely to be most beneficial for *sonic boom attenuation* through the weakening and/or bending of the shock fronts and for delaying shock/boundary layer separation. An increase of gas temperature by a factor of 9 would decrease the local Mach number by a factor of 3 and the local Reynolds number by a factor of 27. At first glance, these changes would appear capable of lowering some flight regimes from supersonic to subsonic or of prodding boundary layers to retransition from turbulent to laminar. Unfortunately, the nature of such aerodynamic effects is considerably more complicated. Nonetheless, when sufficient heating is directed at locations where strong shocks originate on an aircraft flying supersonically, the strength of the emerging shocks may be sufficiently blunted to decrease sonic boom propagation. Similarly, at transonic flow locations where a shock meets a subsonic heated layer, the shock can be attenuated at its foot point by the formation of a so-called lambda shock that has been known to delay unwanted boundary layer separation. These beneficial effects are still subject to research and evaluation before they can be implemented.

## 10.7 Summary

As shown in this and previous chapters, the *small disturbance equation* properly represents important aspects of ideal flows over slender objects and its results apply to compressible subsonic, transonic, and supersonic flows. Our analytical treatment of hypersonic flows has been relatively more limited because its descriptive equations are more numerous, being complicated by changes of flow-specific heat ratio ($\gamma$) along with non-equilibrium effects, heat generation, and heat transfer conditions.

The Prandtl–Meyer expansion formulas can be applied to supersonic flows across our entire range of supersonic Mach numbers inasmuch as gas cooling keeps the value of $\gamma$ in air constant. Shock conditions may also be solved with the equations of Chapter 8 because in transonic flows they are largely weak and normal and because in hypersonic flows there is enough non-equilibrium through their thin layers to make our Chapter 8 assumptions applicable, although $p$-$\rho$-$T$ values after the shock do change noticeably. Hypersonic flows are especially affected by heated regions after the shock since they change gas temperatures and thus $\gamma$ and may eventually cause gas dissociation and even chemical reactions. Recall that all our compressible-flow relations in Chapter 8 are for flows with constant $\gamma$ and for isentropic conditions except through the shock itself.

In Examples 10.1 and 10.2, we examined attributes of an exact solution to the transonic small perturbation equation. The aft portion at the upper region of a thick supercritical airfoil is shown to resemble the "drooped profile" of Example 10.1 as indicated in Figure 10.2. This is a region of special interest where adverse pressure gradients exist and weak shocks may appear. At such location, the boundary layers can be relatively substantial so there is a high tendency for flow separation unless delayed by high turbulence levels in the free stream and/or special airfoil contours.

As mentioned in the Section 10.1, the given expressions for both transonic and hypersonic flows and the objects designed to fly in these regimes look very different from each other. Furthermore, wings designed to fly specifically in these regimes are not necessarily thin. Any more appropriate presentation of these flow regimes, however, require topics that are beyond the scope of this book.

## Problems

**10.1** Consulting a normal shock table for air, tabulate three values for $(1 - Ma_1)$, $(1 - Ma_2)$, $p_{t2}/p_{t1}$, and $\Delta S/R$ for the range $1.0 < Ma_1 < 1.5$. What can you conclude with respect to transonic shock irreversibilities?

**10.2** Show that the potential function given in Example 10.1 is irrotational.

**10.3** Flat plate is moving hypersonically at $\alpha = 5°$. If location 2 is on the upper plate surface where $Ma_2 = 7.0$, find the free stream velocity for approaching air at 15 km altitude.

**10.4** With the flat plate in Problem 10.3, where location 3 is the lower surface, give properties when the oblique shock that develops when $Ma_\infty = 6.15$ and $\alpha = 5°$. Calculate the value of $(p_3/p_1 - p_2/p_1)$ assuming no change of $\gamma$ across shock in hypersonic flow in air.

**10.5** Consider a hypersonic flow over a single-wedge airfoil with half angle $\delta = 15°$ depicted in Figure P10.5.

At an angle of attack $\alpha = 15°$ the upper surface is parallel to the free stream flow. For a Mach number of 20 compare $c_\ell$ values from oblique shock theory to the Newtonian result. (Hint: apply the same calculations used in Example 10.3).

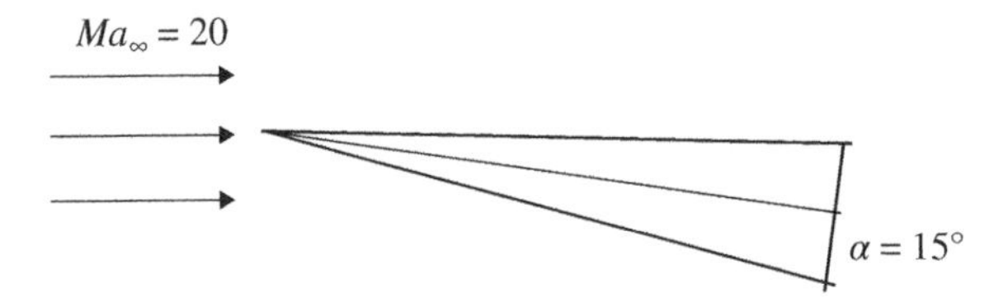


**Figure P10.5**

**10.6** Show that the following equation satisfies the full transonic form as given in Eq. (10.1).

$$\emptyset(x, y) = \frac{V_\infty}{Ma_\infty^2(\gamma + 1)}\left[\frac{x^3}{3y^2} + \left(1 - Ma_\infty^2\right)x\right]$$

## Check Test

**10.1** What is the pressure coefficient $C_p$ on the upper surface of a flat plate in a Mach 7 incoming flow. The angle of attack of the plate is $\alpha = 4.65°$.

**10.2** Why are Prandtl–Meyer flows but not oblique shocks expected to retain their $\gamma$-values in hypersonic flows?

**10.3** For the conditions in Example 10.3, if the flat plate is an angle of attack of $\alpha = 15°$, find the approaching Mach number, the pressure, and the temperature before the oblique shock in air. (Hint: Take =18°)

# 11

# High-Lift Airfoils in Incompressible Flow

## 11.1 Introduction and Approach

In this chapter, we extend thin-airfoil theory to high angles of attack in an attempt to better represent experimental observations. This extension will enable predictions of ideal-flow upper limits such as theoretical values for $c_\ell)_{max}$, which directly affect their three-dimensional counterpart $C_{Lmax}$, as well as enable an enhanced understanding of the role of camber on the coefficients $c_\ell$ and $c_m$. With a more complete incompressible flow foundation, we can better extrapolate results to the full subsonic range by applying the Prandtl–Glauert relation introduced in Chapters 8 and 9. A side bonus of our new formulations is that they eliminate the need to convert angles from degrees to radians when using today's ubiquitous electronic calculators.

The high-lift capabilities of an aircraft depend on its $C_{Lmax}$ and are critically important during takeoff and landing as they directly affect its *stall speed* (see Eq. (6.20) below). In this chapter, we return to purely incompressible flows since they form the foundations for all our subsonic airfoil sectional characteristics. We will examine practical mechanisms that extend the stall condition of airfoils given that with uncambered airfoils, previous schemes to keep the boundary layers attached through appropriately placed blowing and suction ports have resulted in wings of much complexity. To date, a most common approach has been the use of deployable multi-element units that modify the airfoil's cross-sectional shape on demand and thereby the resulting flow pressure profiles. As first discussed in Chapter 5, multi-element wings can be treated as essentially *adjustable camber* airfoils that accomplish shape changes by extending or retracting strategically placed front and rear flaps, although the many gaps created between these surfaces increase drag since such moveable elements are inflexible and overall camber can only be coarsely adjusted. Added leading edge and trailing edge elements remain nested for cruising being only deployed for takeoff and landing (as shown later in Figure 11.4).

$$V_{stall} = \sqrt{\frac{2W}{\rho_\infty S C_{Lmax}}} \tag{6.20}$$

Broadly speaking, high-lift airfoils are attractive for decreasing the land footprint of urban airports and become indispensable for airports serving small mountainous communities as well as at sea on aircraft carriers and on other airstrips with limited surface area. High-speed aircraft and in particular supersonic craft with fixed wings require much smaller wing planforms for cruising compared to subsonic aircraft (see Chapters 9 and 10); so they must operate at high angles of attack before they can reach their final Mach numbers. Furthermore, there has been some interest in aircraft cruising with wings at higher lift coefficient because here *higher payloads* become

*Elements of Aerodynamics: A Concise Introduction to Physical Concepts*, First Edition. Oscar Biblarz.

Companion website: www.wiley.com/go/elementsofaerodynamics

possible at the same speed and angle of attack provided that any added drag (including induced drag) remains tolerable. Also relevant is the fact that that an airplane's "maneuverability" during its smallest possible turn radius and fastest possible turn rate also depends on stalling speed. And presently there is considerable interest on "short takeoff and landing" aircraft that use electric propulsion systems (eSTOL) for urban transportation.

Currently, the highest sectional lift coefficient values without powered-lift assist are around 3.0 for small aircraft with cambered airfoils. Commercial aircraft using high flap defections may also achieve $C_{Lmax} = 3.0$. As will be shown in this chapter, under ideal-flow conditions, the two-dimensional lift coefficient of cambered airfoils peaks around $2\pi$ or about twice currently attainable values suggesting that there is room for improvement. Several techniques are under development based on wing sections whose camber can be morphed by bending a "skeleton-like core" that produces no gaps and can be smoothly and continuously tuned to any desired task while adding little extra drag. Moreover, powered-lift assisted units coupled with boundary layer control are being tested that are expected to attain values of the lift coefficient as high as 7.0.

## 11.2 Objectives

After completing this chapter successfully, you should be able to:

1) Identify flight conditions under which the linear trigonometric approximations (i.e. $\sin\alpha = \alpha$, $\cos\alpha = 1.0$, and $\tan\alpha = \alpha$) are no longer sufficiently accurate.
2) Using our nonlinear sectional lift coefficient for *thin symmetric airfoils*, Eq. (11.1), determine $c_\ell$ at $\alpha = 90°$ as well as the angle of attack for the maximum theoretical value of $c_\ell$.
3) Discuss the effect of camber on the sectional lift coefficients when either $(dz/dx)_w$ or $\alpha_{\ell 0}$ is given and explain the existence of a limiting value for $c_{\ell max}$ of around $2\pi$.
4) Calculate $c_\ell$ for a plain rear flap deployed by an angle $\eta$ under our nonlinear formulations and its effect on the maximum sectional lift coefficient when boundary layers remain attached.
5) Identify the components of common multi-element airfoils and explain the role of each.
6) (*Optional*) Discuss any apparent merits of present and proposed mechanisms morphing camber to achieve high lift.
7) (*Optional*) Identify the reason why the aerodynamic lift force is more efficient than pure vertical thrust for takeoff and landing, thereby explaining the thrust-lowering advantage of airport runways.
8) Enumerate three lift coefficient airfoil modifications highlighted in this chapter.

## 11.3 Nonlinear Thin Airfoil Theory

*Thin airfoil theory* remains largely unchanged since its introduction in the 1920s. It contains some mathematical advanced detail (e.g., the solution of an integral equation, see Eq (5.7)) and its results properly represent measured sectional airfoil characteristics in the linear range of angles of attack. Furthermore, this theory is a useful application of the principle of superposition of elementary flows and its formulations with respect to $\alpha$ remain accurate even in today's digital computer world. Nonlinear versions of thin airfoil theory that extend results into angles of attack above 20° have been proposed (i.e., Hoerner 1975) providing useful insights that are no more complicated than traditional versions. At present, airfoils with thin and attached boundary layers must operate in very limited ranges of angle of attack and the realization of high-lift airfoils remains technologically challenging.

Using an initial result from thin airfoil theory for symmetric foils, namely Eq. (5.10a),

$$c_\ell = 2\pi \sin \alpha$$

we might conclude that, theoretically, lift coefficients reach a maximum at $\alpha = 90°$. At this orientation, however, a thin flat plate has no surface to support any finite vertical force even if by some means we could generate the needed bound vorticity at such orientation. From a geometrical perspective, at any airfoil angle of attack the *portion of the airfoil which transfers lift to the vehicle is not the chord itself but its projection* on a plane that the velocity vector makes with the airfoil's span thus introducing a cos$\alpha$-multiplier into the equation above. This interpretation is depicted in Figure 11.1 showing a thin cambered airfoil.

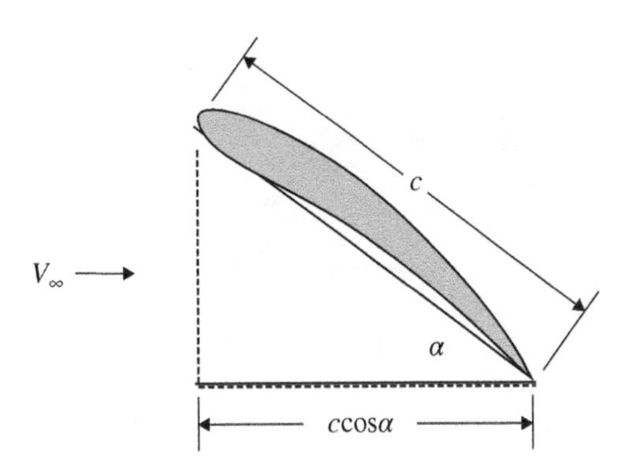


**Figure 11.1** Chord projection defining the effective planform area in the lift direction.

The Kutta–Joukowski aerodynamic force results from a well-defined interplay between the circulation bound to an airfoil and the free stream flow velocity approaching it. It was originally derived for spinning cylinders immersed in uniform flows. Unlike airfoils, with cylinders, there is no angle of attack variation of their *lifting surface* (which equals the chord line in two-dimensional flows) because its radius does not change in magnitude with cylinder position during rotation. Symmetric airfoils, however, only generate circulation at angle of attack which geometrically modifies their lifting surface area. A rigid but thin flat plate has no circulation at $\alpha = 0°$ so that there is zero lift. However, at $\alpha = 90°$, where circulation according to airfoil theory is a maximum, there is also no lift since the lifting-surface projection of has vanished. The drag coefficient is not affected by such angle variations and remains a function of the full chord because skin drag originates in the enveloping boundary layers. Recall that in Chapter 7, we introduced the Reynolds number dependance of the drag based on the entire chord length.

### 11.3.1 Thin Symmetric Airfoils

For *thin symmetric airfoil configurations*, we proceed next to write Eq. (5.9) as

$$\ell = \rho_\infty V_\infty \sin \alpha \int_0^c \gamma_s(x_o) dx_o = \pi \rho_\infty V_\infty^2 c \sin \alpha \cos \alpha \tag{5.9mod}$$

As formulated here, the modification of the upper integral limit in Eq. (5.9) accounts for a changing *lifting planform area*, which was originally the entire chord ($c$) and now becomes $c(\cos \alpha)$ to represent the projection of the wing's sectional area on the "lifting-plane." With this change and for angles beyond $\alpha = 15°$, the resulting ideal lift coefficient and related parameters for symmetrical airfoils can now be written as

$$\begin{aligned} &c_\ell = 2\pi \sin \alpha \cos \alpha = \pi \sin (2\alpha) \\ &\alpha_{\ell 0} = 0 \quad \text{and} \quad c_{mac} = 0 \quad \text{with} \quad x_{ac} = c/4 \\ &c_{\ell\, max} = \pi \quad \text{at} \quad \alpha_{max} = 45° \end{aligned} \tag{11.1}$$

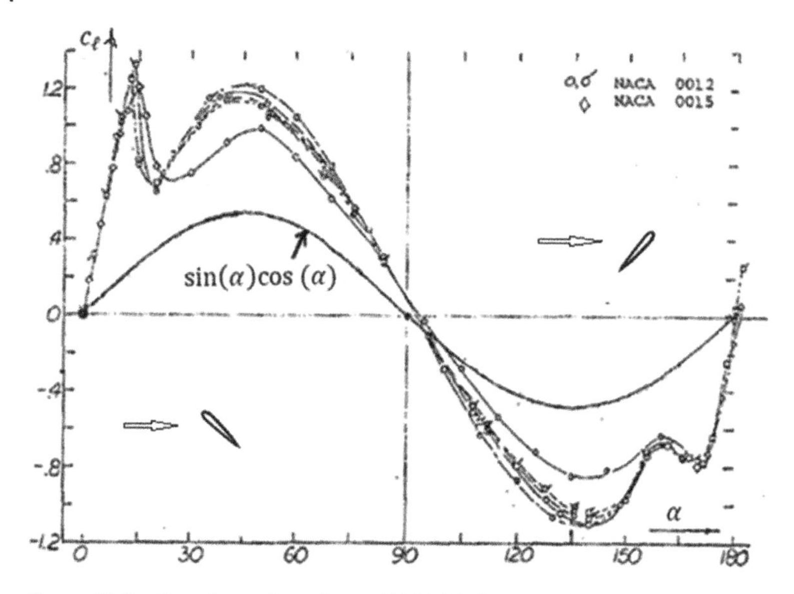


**Figure 11.2** Experimental results on NACA 0012 and 0015 symmetric airfoils used for helicopter rotors from Hoerner (1975). The product sin($\alpha$)cos($\alpha$) is also displayed.

Note in the top equation in (11.1), after the second equal sign, that having extended the curve beyond its linear range approximation, the angle of attack is doubled inside the sine-function (using $\sin(2\alpha) = 2 \sin \alpha \cos \alpha$). More importantly, when the boundary layers surrounding the airfoil can kept thin and attached, this formulation leads to $c_{\ell max} = \pi$ at $\alpha = 45°$and not $2\pi$ at 90°. The set of formulas presented as Eq. (11.1) should be more correct than those given in Chapter 5 – published data (see Abbott and von Doenhoff 1949; Hoerner 1975) reveal that no measurements on symmetric airfoils are known to exceed $c_{\ell max} = 3.14$, with most symmetric airfoils typically peaking below $c_\ell = 2.0$ at $\alpha < 20°$. Experimental results on two classic NACA airfoils with designators 0012 and 0015 are shown in Figure 11.2 for an extended range in $\alpha$.

Figure 11.2 merits the following interpretation: Both airfoils tested are classic NACA designs. During the wind tunnel experiments each airfoil's angle of attack was rotated from 0° to 180° (at several different Reynolds numbers). The measured $c_\ell$-curves do not flip over exactly at $\alpha =$ 90° as the product sin($\alpha$)cos($\alpha$) does because, in real airfoils, some lift is supported by their small thickness. Moreover, mirror images of subsonic symmetric airfoils must differ because their rounded nose and sharp trailing edge becomes a sharp nose and rounded trailing edge so that after $\alpha = 90°$ the flow around the tested airfoils, having beed reversed, generates different lift and boundary layer patterns. Nontheless, some remarkable behavior is apparent since at both $\alpha$-ends in Figure 11.2 $c_\ell$ closely follows the magnitude of $\pi \sin(2\alpha)$ reflecting the presence of attached boundary layers. Measurements between those two ends peak at 45° and only slighltly beyond 135° – very nearly equal to the sin($\alpha$)cos($\alpha$) $\alpha$-peaks. These data also show that, even after boundary layers begin to detach, $c_\ell$ recovers most of its magnitude at the higher angles of attack (with different extrema) thus reflecting some proportionality of $c_\ell$ to the product sin($\alpha$)cos($\alpha$). Helicopter airfoils have high rates of change of pitch so flows are unsteady which is known to lessen the amout of boundary layer detachment during each cycle. The most significant takeaway from Figure 11.2 is that, for symmetric thin airfoils, the Eqs. in (11.1) better display the overall dependence of $c_\ell$ on the complete $\alpha$-range than their linear counterparts.

Hoerner (1975) provides the following form for applicable lift ($c_\ell$) and for the sectional drag ($c_d$) coefficients at high angles of attack:

$$c_\ell = (1.8 \text{ to } 2.0) \sin\alpha \cos\alpha \tag{11.2a}$$

$$c_d = (1.8 \text{ to } 2.0)\sin^2\alpha \tag{11.2b}$$

### 11.3.2 Thin Cambered Airfoils

We proceed with modification to the $c_\ell$ and $c_{\ell\alpha}$ thin airfoil theory formulations from Chapter 5 to represent higher-angles of attack in *cambered airfoils*. We also include a modified version of the sectional zero-lift angle $\alpha_{\ell 0}$ which introduces the dependence on the slope of the camber line $(dz/dx)_w$. As discussed in Chapter 5, using Fourier series expansion coefficients from the general circulation distribution in two-dimensional airfoils requires changing the Cartesian coordinate to polar form, namely, $x \equiv 0.5c(1 - \cos\theta)$. We thereby obtain the following results for large magnitudes of $\alpha$:

$$c_\ell = \pi \sin(2\alpha) + 2\cos^2\alpha \int_0^\pi \left(\frac{dz}{dx}\right)_w (\cos\theta - 1)d\theta \tag{11.3}$$

$$c_{\ell\alpha} = 2\pi\cos(2\alpha) - 2\sin(2\alpha)\int_0^\pi \left(\frac{dz}{dx}\right)_w (\cos\theta - 1)d\theta \tag{11.4}$$

$$\alpha_{\ell 0} = \tan^{-1}\left[-\frac{1}{\pi}\int_0^\pi \left(\frac{dz}{dx}\right)_w (\cos\theta - 1)d\theta\right] \tag{11.5}$$

These formulations allow finding of the angle of attack for maximum lift in cambered airfoils by setting the slope $c_{\ell\alpha} = 0$.

$$\alpha_{max} = \frac{1}{2}\tan^{-1}\left[\frac{1}{\pi}\int_0^\pi \left(\frac{dz}{dx}\right)_w (\cos\theta - 1)d\theta\right]^{-1} \tag{11.6}$$

With camber restricted to thin airfoils, values of $\alpha)_{max}$ may vary between 30° and 45°, and $\alpha_{\ell 0}$ between 0° and − 30°. The relationship between $\alpha_{max}$ and $\alpha_{\ell 0}$ is useful and may be written as

$$(\tan(2\alpha_{max}))(\tan(-\alpha_{\ell 0})) = 1.0 \tag{11.7}$$

Equations (11.3) and (11.4) may now be written in a form equivalent to Eq. (5.17) in terms of $\alpha_{\ell 0}$ as (recall $\alpha_{\ell 0} < 0°$),

$$c_\ell = 2\pi\cos\alpha(\sin\alpha - (\cos\alpha)(\tan\alpha_{\ell 0})) \tag{11.8a}$$

$$c_{\ell\alpha} = 2\pi\cos(2\alpha) + 2\pi[(\sin(2\alpha))(\tan(\alpha_{\ell 0}))] \tag{11.8b}$$

**Example 11.1** In Example 5.2, we looked at camber in a *circular-arc airfoil* which is made up from the arc of a circle of radius $r_0$, chord c, and rise $\kappa c$, where $\kappa$ is a small constant; this profile provides a useful way to calculate increases from the camber on the lift coefficient by changing the value of $\kappa$. Here, we want to look at resulting values of $\alpha_{\ell 0}$ and $\alpha_{max}$ as functions of $\kappa$ with the formulations introduced in equations (11.5) and (11.6).

$(dz/dx)_w = 4\kappa\cos\theta \quad 0 \le \theta \le \pi$ so that the value of the integral in equation (11.5) becomes $(\pi/2)(4\kappa)$.

$$\alpha_{\ell 0} = \tan^{-1}\left[-\frac{4\kappa}{\pi}\int_0^\pi \cos\theta(\cos\theta - 1)d\theta\right] = \tan^{-1}(-2\kappa)$$

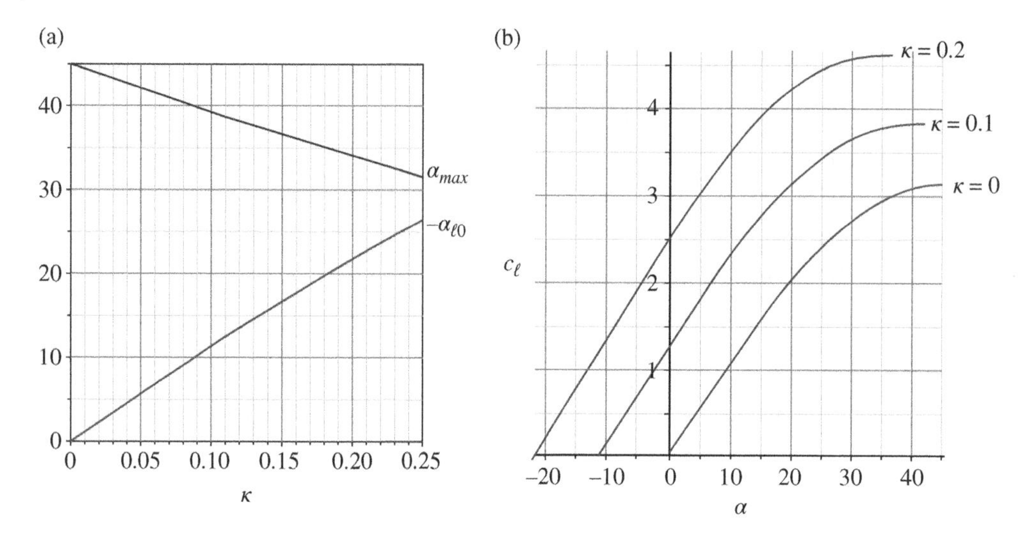


**Figure E11.1** Graph (a) shows $\alpha_{max}$ and $-\alpha_{\ell 0}$ vs $\kappa$, both angles in degrees. Graph (b) $c_\ell$ vs $\alpha$ for $\kappa$ = 0, 0.1, and 0.2, all $\alpha$s are in degrees. These curves represent strictly theoretical values without stall conditions.

$$\alpha_{max} = \frac{1}{2}\tan^{-1}\left[\frac{4\kappa}{\pi}\int_0^\pi \cos\theta(\cos\theta - 1)d\theta\right]^{-1} = \frac{1}{2}\tan^{-1}\left(\frac{1}{2\kappa}\right)$$

The plots of $-\alpha_{\ell 0}$ and $\alpha_{max}$ vs $\kappa$ are shown in Figure E11.1a. Figure E11.1b shows $c_\ell$ vs $\alpha$ for three values of $\kappa$.

The anticipated behavior for $\alpha_{max}$ and $-\alpha_{\ell 0}$ is clearly seen in Figure E11.1a and verifies Eq. (11.7). And as evident in Figure E11.1b all curves go through a maximum in $c_\ell$ and are displaced to the left as $\alpha_{\ell 0}$ increases in magnitude with increasing $\kappa$. As camber increases the numerical value of $c_{\ell max}$ also rises beyond $\pi$ which represents the limit for symmetric airfoils.

The overall trends indicated in Example 11.1 as shown in Figure E11.1a and b are considered appropriate. They indicate the angle $\alpha_{max}$ receding and its companion $\alpha_{\ell 0}$ becoming more negative with increasing amounts of camber and this is seen in experimental results. The topmost value of each lift coefficient ($c_\ell$) shown increases with increasing camber and is higher than available data because the boundary layer would typically detach before any theoretical value of $\alpha_{max}$ can be reached (depending on the amount of boundary layer control present). It is insightful, however, to look for an ultimate maximum $c_\ell$ predicted with for cambered airfoils by Eq. (11.8a) and see how it compares to previous theoretical results from Chapter 4. Because in *circular-arc airfoils* the maximum value of the constant $\kappa$ is 0.5 (i.e. it cannot exceed the radius of a circle located at its chord), we see from Figure 11.3 that the highest contribution from camber to the lift coefficient at $\alpha = 0°$ as given with this "hemi-circular cylinder airfoil" amounts to $2\pi$. This is significant because the theoretical maximum lift on a rotating *full circular cylinder* is $4\pi$ and here we have modelled one-half of a cylinder under maximum circulation. Additional lift from the airfoil at angle-of-attack enhances this value to $c_{\ell max} = 7.6$ at 22.8°, but when $\kappa c$ equals the cylinder radius, we have exceeded the required *small camber* approximation which brings into question the legitimacy of values other than at $\alpha = 0°$ in Figure 11.3.

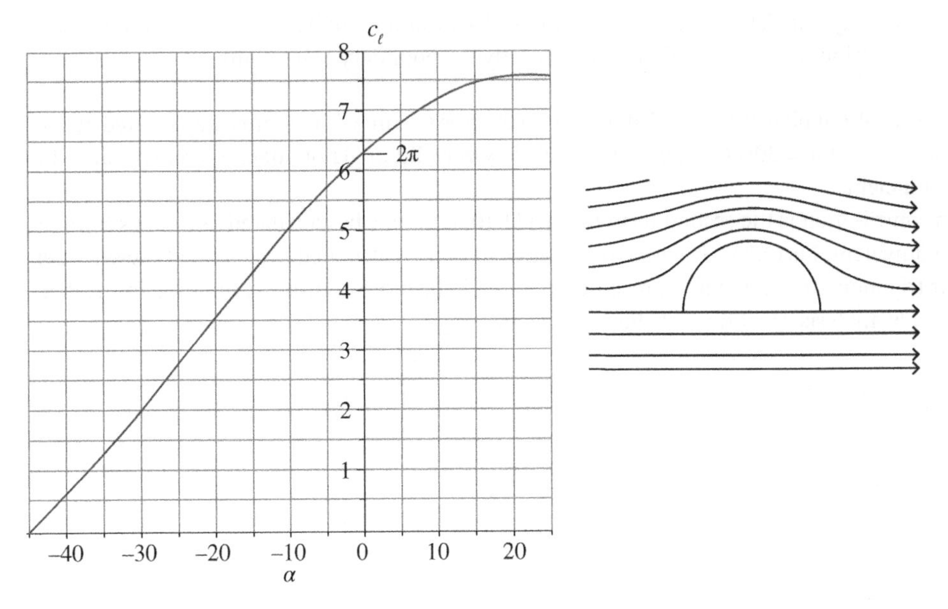


**Figure 11.3** Theoretical $c_\ell$ vs $\alpha$ when the camber line becomes one-half of a cylindrical profile. The sketch detailed on the right shows ideal streamlines for $\alpha$ = 0° with the bottom surface of the semi-cylinder close to the flow as in a Quonset hut suspended in a moving stream.

### 11.3.3 Flaps on Symmetric Airfoils at High $\alpha$

We will assume that flaps, slats, and other high-lift devices continuously modify camber without adding major loses as the angle of attack exceeds its linear range and that the airfoil may be replaced by three additive parts as in Figure 5.1. Equation (5.15) may be used to find effects of a single flap on a flat plate deployed by the angle $\eta$ with the flap located at $x_h < c$ so that $\theta_h = \cos^{-1}(1 - 2x_h/c)$. In order to apply Eq. (5.15), we note that the aft-portion of the plate is turned as in Figure 5.10 so that $dz/dx)_w = \tan(-\eta) = -\tan(\eta)$ which is constant for any given flap deflection. All resulting enclosed angles in Figure 5.10 must remain small and amenable to the usual trigonometric approximations. As written in Chapter 5, Eq. (5.23) for $\Delta\alpha_{\ell f0}$ already represents values of $\eta > 15°$ in Cartesian coordinates. Equations (11.9) and (11.10) show resulting forms for $c_\ell$ and $\Delta c_{mc/4}$ but in the usual polar coordinates (see Eq. (5.14)).

$$c_\ell = \pi \sin(2\alpha) + 2\int_{\theta_h}^{\pi} \left(\frac{dz}{dx}\right)_w (\cos\theta - 1)d\theta \tag{11.9}$$

$$c_\ell = \pi \sin(2\alpha) + 2[\sin\theta_h + (\pi - \theta_h)]\tan(\eta) \tag{11.10}$$

$$\text{and} \qquad \Delta c_{mc/4} = \frac{1}{2}[\sin\theta_h(\cos\theta_h - 1)]\tan(\eta) \tag{11.11}$$

Comparison of results from these formulations with experiments shows only moderate agreement partly because trailing edge flaps are often fully immersed in the airfoil's boundary layers. The equations for straight trailing edge flaps may also be applied straight *leading edge slats* by using an appropriate displacement angle $\eta$ in the opposite direction and hinged at a proper forward $\theta_h$ or $x$-location. Moreover, we may calculate the resulting action of both a straight flap and a straight slat deployed on the same airfoil as long as the thin airfoil assumptions of Chapter 5 are not greatly exceeded. As already stated, because of their leading edge location, camber changes from slats make insignificant contributions to $c_\ell$ so their use is strictly for boundary layer control.

**Example 11.2** A flat plate is oriented at $\alpha = 20°$ to the free stream with a simple flap located at $0.8c$ which is deflected at $\eta = 30°$. Calculate the ideal-flow zero-lift angle-of-attack increment and the total lift coefficient.

The flap contribution is the second term in Eq. (11.10) and may be calculated as follows from its Cartesian equivalent, Eq. (5.23). The total lift coefficient is also calculated. Note that at these higher angles, our trigonometric function representations should be more accurate (even though the deficiencies of this analysis mentioned in Figure 5.11 still apply).

$$\Delta\alpha_{\ell f0} = \left[\left(1 - \frac{\cos^{-1}(1 - 2x_h/c)}{\pi}\right) + \frac{2\sqrt{x_h/c - (x_h/c)^2}}{\pi}\right]\tan(\eta)$$

$$= \left[\left(1 - \frac{\cos^{-1}(1 - 1.6)}{\pi}\right) + \frac{2\sqrt{0.8 - (0.8)^2}}{\pi}\right]\tan(30) = 0.3174 = 18.19°$$

$$\Delta c_\ell = 2\pi\Delta\alpha_{\ell f0} = 1.90$$

$$c_\ell = \pi\sin(40°) + 1.90 = 3.92$$

## 11.4 Pitching Moment at c/4 and the Aerodynamic Center

The *aerodynamic center* was defined as that location along the airfoil's chord where the pitching-moment coefficient is independent of angle of attack $\alpha$. At higher values of this angle, we may also use Eq. (5.19) for the pitching moment about the quarter-chord location but cambered airfoils now develop a slight dependence on the angle of attack.

$$c_{mc/4} = \frac{\cos^2\alpha}{2}\int_0^\pi \left(\frac{dz}{dx}\right)_w (\cos(2\theta) - \cos\theta)d\theta \tag{11.12}$$

Experiments with cambered airfoils below stall indicate that the aerodynamic center stays close to the $c/4$ location, varying between 23 and 27% chord. In symmetric airfoils, the slope of the camber line is zero $[(\frac{dz}{dx})_w = 0]$, i.e. $c_{mc/4} = c_{mac} = 0$. For a flapped symmetric airfoil at an angle $\eta$, the flap contribution to the moment is shown as Eq. (11.11).

The dependence of Eq. (11.12) on angle of attack remains small. For example, at $\alpha = 18°$ the magnitude of $\cos^2\alpha = 0.90$ in Eq. (11.12) so that we are dealing with about a 10% discrepancy for ordinary angles of attack (however at $\alpha = 26.4°$ $\cos^2\alpha$ equals 0.80 becoming more pronounced). Abbott and von Doenhoff (1949) extensively reports on experimental results on the aerodynamic centers and $c_{mc/4}$ for NACA airfoils.

**Example 11.3** For the circular-arc airfoil of Example 11.1 show that $c_{mc/4} = -\pi\kappa\cos^2(\alpha)$. Plot this function for $\kappa = 0.1$ and 0.2.

When we input $(dz/dx)_w = 4\kappa\cos\theta$ into Eq. (11.12), the resulting integral becomes $-2\pi\kappa\cos^2(\alpha)/2$ which yields the desired answer.

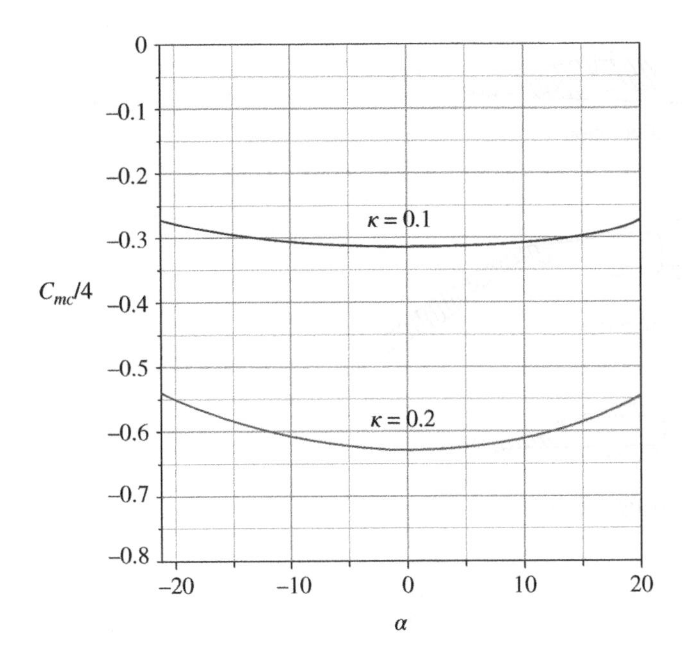


**Figure E11.3** Theoretical values on $c_{mc/4}$ for $\kappa = 0.1$ and 0.2. The camber line symmetry about the mid-chord is reflected in the trough-like appearance of $c_{mc/4}$.

$$c_{mc/4} = \frac{\cos^2(\alpha)}{2}\int_0^{\pi} 4\kappa\cos\theta(\cos(2\theta) - \cos(\theta))d\theta = -\pi\kappa\cos^2(\alpha)$$

Equation (11.12) makes the $c/4$-location of the aerodynamic center dependent on the angle of attack and thus not a true aerodynamic center but as we stated for ordinary angles, $\cos^2\alpha$ is no more than 10% off. As seen in Figure E11.3, the slight angle dependence shows as concave about $\alpha = 0°$ because the circular-arc airfoil has camber symmetry about $c/2$ – a behavior that is also observed in NACA airfoil data when the camber line is similarly distributed (with sharp changes in magnitude as the airfoil stalls).

## 11.5 High-Lift Wing Mechanisms

Conventional symmetric as well as cambered airfoils without high-lift components attain $c_\ell$-values well below any theoretical maximum predictions. On-demand variable-camber airfoils, however, are capable of higher values of $c_{\ell max}$ from both the camber contribution and delays of airfoil stall. Having already discussed single trailing-edge flaps in Section 5.7 and in Section 11.3, we continue here with experimental results from a multi-element airfoil configuration that includes leading edge slats and multiple rear flaps as shown in Figure 11.4. These components are designed to increase camber and extend the useful $\alpha$-range before stall by managing the boundary layers and producing noticeably higher values of $c_{\ell max}$. There are many multi-element airfoil configuration designs but all consist of components that can remain folded or nested during cruise conditions. These separate but rigid flap elements are designed to adjust camber at wing locations where they can be most effective. As discussed in Chapter 5, rear-end flaps produce camber enhancements most efficiently achieving high values of $\alpha_{\ell 0}$, whereas front-edge slats do not

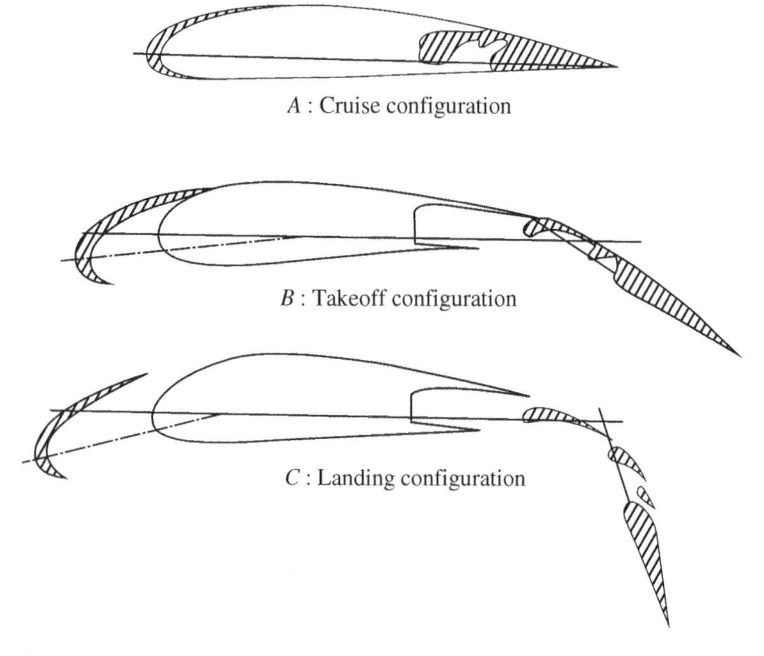


**Figure 11.4** Multi-element airfoil. Notice the increases of the actual chord length between the top or fully nested configuration and the other two. *Source:* From Anderson (2017) (additional figures may be found searching the Web for "airfoil edge devices images").

measurably change $\alpha_{\ell 0}$. Some early types of high-lift designs also incorporated *jet flaps* or used internally compressed gases injected at selected rear airfoil sections to energize the boundary layers.

As seen in Figure 11.4, hardware used during high-lift flight operations adds high camber to the airfoil but the gaps between their multi-element parts also add drag and each deployable component has a fixed shape. Another complication that multi-element airfoils introduce to our analysis concerns an airfoil chord enlargement from their cruise or nested configuration. As airfoil areas are customarily rated by their nested chord configuration, at the angles of attack associated with takeoff and landing, this smaller chord value yields artificially increased lift coefficients from the data. Therefore, lift coefficients based on a nested chord need to be corrected whenever wing multi-element are deployed. To account for increases in chord-length with deployed flaps and slats, we modify the lift coefficient as follows:

$$c_\ell = c_{\ell 0}\left(\frac{c_0}{c}\right) \tag{11.13}$$

In this relation, $c_{\ell 0}$ is the lift coefficient using $c_0$ for the nested cruise configuration chord value, and $c_\ell$ the actual coefficient using $c$, the fully deployed chord length. For total or $C_L$ coefficient values, an additional correction is needed because most flap units do not cover the entire wing span (e.g. see Figure 7.8).

Next, we examine the experimental sectional values shown in Figure 11.5 for the NACA $64_1$-012 configuration which has high-lift appendages that can be activated independently. For this symmetric airfoil, we first consider the effects from the leading edge flap or slat because its movement does not significantly affect the camber. Experimental results of the nested airfoil give $c_\ell)_{max} = 1.42$ (the dotted circles), whereas with the leading edge fully extended it reaches $c_\ell)_{max} = 1.85$ (the dotted squares). The slat extends the chord ($c_0$) by $0.10c$ allowing the angle of attack to reach $\alpha)_{max} = 18°$. The span corrected chord from Eq. (11.13) lowers this to $c_\ell)_{max} = 1.68$. Using Eq. (11.1) we calculate

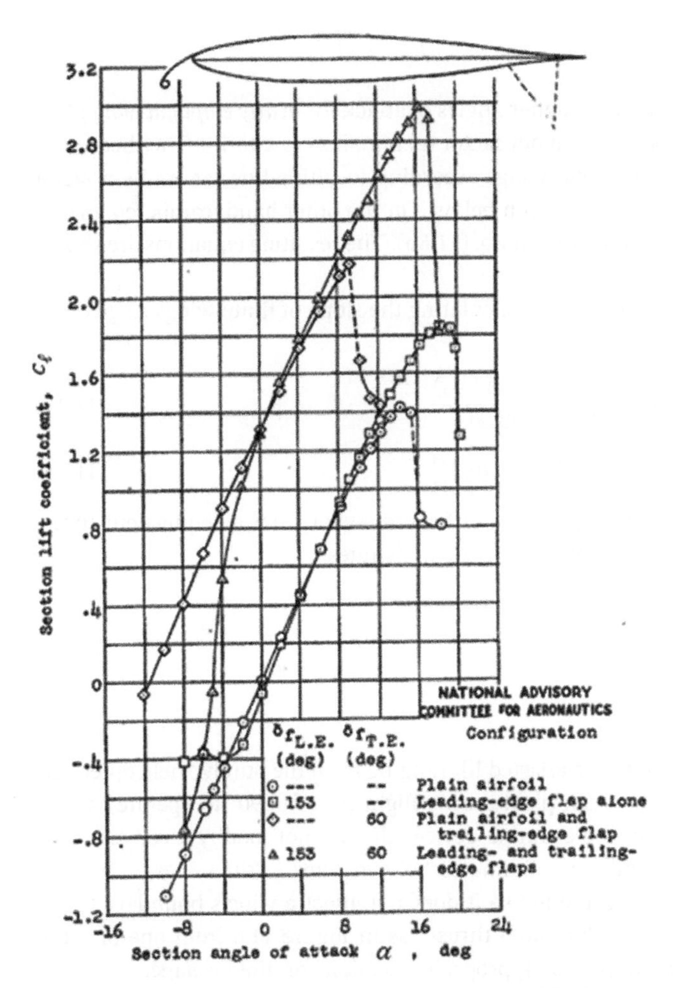


**Figure 11.5** Section lift characteristics for the NACA $64_1$-012 section equipped with a 0.10*c* upper-surface leading edge flap alone, and in combination with a 0.2*c* trailing edge split flap ($Re = 6 \times 10^6$). *Source:* From NACA-TN 1277, May 1947.

$$c_\ell = \pi \sin(2\alpha) = 1.85$$

which exceeds the corrected value above and the discrepancy likely comes from having used the ideal slope ($2\pi$) instead of a more realistic lower value (see Section 11.7.2). Finally, we note that at $\alpha = 18°$, the linear product $2\pi\alpha = 1.97$ or about 6.5% higher than 1.85 and 17% higher than 1.68.

Additional data shown with dotted triangles in Figure 11.5, from a combination of leading and trailing edge flaps, starts below the diamonds but eventually overtakes them and here the maximum lift coefficient becomes 67% higher than the unflapped airfoil. These data appear highly nonlinear up to $\alpha = 0°$ and linear beyond that, to $\alpha = 16°$. We note that the rear flap contribution calculated from Eq. (11.7) can significantly overestimate the total lift value at $\eta = 60°$ (as already mentioned in Section 5.7). Experimental stall values at $\alpha = 16°$ and 18° define a $c_{\ell max}$ that reflects the boundary layers suddenly detaching, but both experimental curves appear linear just before this. In summary, while the *trends* represented by Eqs. (11.8a) and (11.9) are reflected by the data in Figure 11.5, our calculated $c_{\ell max}$ with and without flaps does overestimate experimental values and this is primarily due to real flow effects.

## 11.6 Finite Wings

We extend here formulations in Chapter 6 to higher angles of attack for *nearly elliptical wing planforms* with the expectation that their span efficiency factor ($e$) remains close to 1.0. The theoretical two-dimensional slope ($2\pi$) decreases in finite wings according to lifting-line theory by a factor related to the aspect ratio (***AR***) of the wing as given below. On the other hand, cambered airfoils slightly enhance this slope as may be surmised from Eq. (11.8b). The resulting equations are shown next as (11.14) and (11.15).

Replacing $2\pi$ by $c_{\ell\alpha}$ in Eqs. (11.4) and (11.5) for calculating the slope of finite wings as given by Eq. (6.17) and using Eq. (11.8a), we get

$$C_{L\alpha} = \frac{c_{\ell\alpha}}{1 + \frac{c_{\ell\alpha}}{\pi e AR}} = \frac{2\pi\cos(2\alpha) + 2\pi\sin(2\alpha)\tan(\alpha_{\ell 0})}{1 + \frac{2\cos(2\alpha) + 2\sin(2\alpha)\tan(\alpha_{\ell 0})}{eAR}} \tag{11.14}$$

$$C_L = C_{L\alpha}\left[(\cos(\alpha))(\sin(\alpha)) - (\cos^2(\alpha))(\tan(\alpha_{\ell 0}))\right] \tag{11.15}$$

Note that even at $\alpha = 10°$ results from Eq. (11.15) differ somewhat from its equivalent three-dimensional counterpart in Chapter 6 as Problem 11.7 demonstrates.

## 11.7 Enrichment Topics

### 11.7.1 Power-Assisted Lift

There are several different methods for power-assisted lift ranging from the purely "helicopter vertical flight" mode to propulsion units that can pivot during flight by a full 90° to operate as conventional thrust engines. In this section, we examine devices that do not modify a conventional wing's lift-producing characteristics, and this eliminates "blown flaps," "tilt wings," and "tilt rotors" – in our treatment; jet thrust is positioned so it does not affect a wing's boundary layers or acts as jet or blown flaps. We examine "vectored thrust" as in Figure 11.6 from one or more engines suitably mounted under the wings or at appropriate locations on the fuselage.

While operating in lift-assisted mode, the engine thrust pivots so that the total lift ($L$) equals or may slightly exceed the weight ($W_0$) during takeoff or landing. Such engines may move through an angle ($\beta$) – where $\beta = 0°$ for takeoff or landing and for cruising $\beta$ pivots to 90°. An intermediate position is depicted in the Figure 11.6.

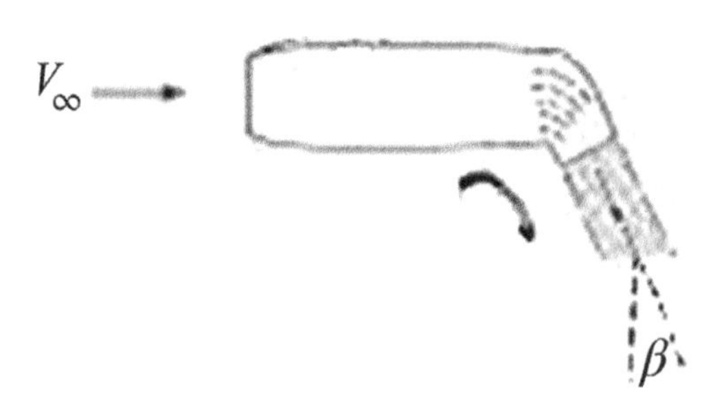


**Figure 11.6** Lift assist technique with vectored thrust in the body of aircraft. The center of gravity of the aircraft needs to be factored in when positioning several lift-vector assist units.

During quasi-steady, level, unaccelerated flight, there are no net forces (since inertial effects have vanished) and we can easily estimate the ratio of total thrust to weight ($T/W_0$) as well as the ratio of the free stream velocity to its end value at cruising ($V_\infty/V_{\infty max}$). We know that during hovering in purely vertical takeoff or landing $T = W_0$ and the horizontal speed $V_\infty$ or $q_\infty = 0$. Moreover, during steady-state cruising $V_\infty = V_{\infty max}$ with $T/W_0 < 1.0$ since *aerodynamic lift* can be more *efficient* than the *raw thrust* that generates it. Also, as the aircraft transitions to its cruise mode overall drag coefficients decrease as flaps and slats can then be stowed and the need

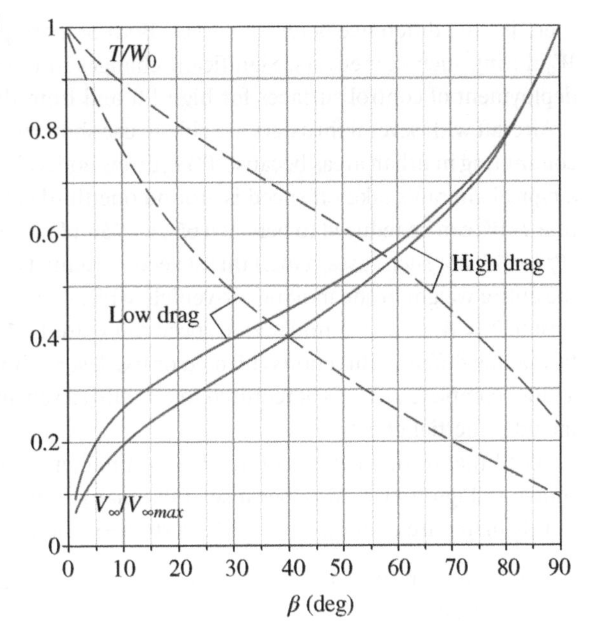


**Figure 11.7** Composite plot of $V_\infty/V_{\infty max}$ (solid lines) and $T/W_0$ (dashed lines) as a function of engine exhaust pivot angle $\beta$ (deg) excluding transient conditions. During takeoff $\beta = 0° \rightarrow 90°$, whereas during landing $\beta = 90° \rightarrow 0°$.

for high-lift coefficients vanishes enabling the drag coefficient to decrease. For our model, we assume for expediency that lift and drag are related as during operation at $L/D)_{max}$, where $c_d = C_{Di}$, see Eq. (6.18). In our calculations, we have used $e\boldsymbol{AR} = 8.6$. The following equations summarize our approach and Figure 11.7 depicts results ($\beta$ is in degrees).

$$W_0 = T\cos\beta + q_\infty SC_L \tag{11.16}$$

$$D = T\sin\beta \equiv 2q_\infty SC_{Di} = \frac{2q_\infty SC_L^2}{e\pi\boldsymbol{AR}} \tag{11.17}$$

$$\frac{V_\infty}{V_{\infty\, max}} = q_\infty\left[\frac{2C_L^2\cot\beta}{e\pi\boldsymbol{AR}} + C_L\right] = q_{\infty\, max}C_{Lmin}$$

$$\frac{V_\infty}{V_{\infty\, max}} = \sqrt{\frac{C_{Lmin}/C_L}{\left(\frac{2C_L}{e\pi\boldsymbol{AR}}\right)\cot\beta + 1.0}} \tag{11.18}$$

$$\frac{T}{W_0} = \frac{1}{cos\beta + \frac{e\pi\boldsymbol{AR}}{2C_L}sin\beta}$$

$$C_L \approx 2.0 - 0.021\beta$$

$$\frac{C_{Lmin}}{C_L} \approx 0.2/(2.0 - 0.021\beta)$$

Figure 11.7 shows results in terms of non-dimensional ratios which only depend on the pivot angle under our quasi-steady state conditions. For any value of $V_\infty/V_{\infty max}$, this figure shows the least amount of thrust needed to keep the vehicle airborne along with the thrust angle required under both low and high drag conditions. Unsteady trajectories are expected to proceed from high-drag locations during takeoff and from low drag locations during landing. Takeoff would begin at the left side of the figure and end at the right with thrust varying from high drag to low drag. Landing would start on the right side low-drag curve and terminate on the left where there is little difference between curves. Both of these procedures assume negligible variations in $W_0$ during such maneuvers. Significant changes in the drag coefficients arise primarily from the deployment of control surfaces for high lift and from the *ground effect* (discussed in Chapter 6).

Aircraft with several thrusters are able to use short runways for takeoff which is advantageous for commuting in urban areas because $T/W_0$ drops noticeably when $\beta > 0°$. For example, assuming that a typical aircraft's take-off speed is around one-third (1/3) of its cruise speed, Figure 11.7 indicates that $T/W_0$ can be reduced to between 65 and 75% of the vertical take-off value with $\beta$'s ranging from 17° to 27°. For such cases, a total thrust vector oriented between these angle locations is able to carry the entire weight of the moving air-vehicle while compensating for its drag with $T < 0.8W_0$ in contrast to $T = W_0$ at $\beta = 0°$ under zero speed in Figure 11.7. When individual thrust units can exhaust flow along different directions (pivoted or fixed, some horizontally and some vertically positioned), a more complete analysis is required to establish a schedule between individual thrusters, one that includes inertial effects.

Recall that during periods of acceleration higher thrust-to-weight ratios would be needed than shown in Figure 11.7. Also, because of the many simplifications in the analysis, the curves shown in this figure are only qualitative. Nevertheless, the following observations are relevant:

a) The velocity ratio representation is nonlinear but largely diagonal with only minor effects from drag variation.
b) Thrust-to-weight ratios vary noticeably within the assumed drag variations.
c) The reduced-thrust advantage from aerodynamic lift over powered lift suggest use of short runways as more optimal than the purely vertical flight stages (i.e. short-take-off-and-landing or "STOL" vehicles).

### 11.7.2 Comparison of the $\alpha$-Dependence in Eqs. (5.17), (11.8a), and (11.12) with Data

Continuing the comparisons started in Section 11.5 of our formulations with data, we examine here the $\alpha$-dependence of Eq. (11.8a) and its more approximate form, Eq. (5.17), and with Eq. (11.12) we compare data values for $\alpha$ in a range of most NACA wing-section data ($\pm 15°$); all data points come from the NACA 4415 wing section which is representative of many such cambered airfoils. Figure 11.8 is a plot of these three curves, where $\alpha_{\ell 0} = -4°$ and $c_{mc/4} = -0.085$ at $\alpha = 0°$. The difference between Eqs. (11.8a) and (5.17), apparent above $\alpha = 8°$, represents two errors, the first comes from the linear approximation of the trigonometric functions the other from keeping the airfoil's *lifting-surface* as the entire chord length at all angles of attack. For $c_{mc/4}$, the scale in Figure 11.8, has been magnified four times and these data show more than the slight upward concavity that arises from the $\cos^2\alpha$-term in Eq. (11.12). We may surmise from this figure that mid-range data magnitudes are well represented by all three equations when the lift slope is decreased to 93%-theoretical, and that our nonlinear behavior representation is better but begins to over-represent experimental values after $\alpha = \pm 12°$ likely because of the stall-condition onset reflecting nonideal behavior.

### 11.7.3 Unsteady Flows

Some lift advantages can be obtained from unsteady flows created with fast oscillating and pivoting wings, even in turbulent air flows. Beyond certain dynamic effects from *leading edge vortices* over their wings, bumblebees, hummingbirds, and bats, all benefit from features of unsteady flows that enable the boundary layers to remain attached on their wings at relatively high angles of attack thereby achieving higher lift forces before wing-stall. This allows for greater loads or alternatively smaller wing areas to achieve the same weight-lifting or hover capability because, for short enough

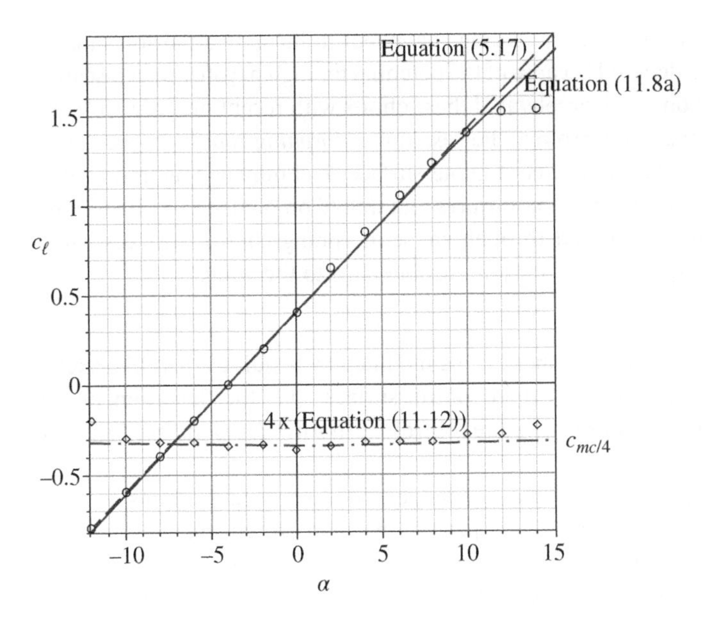


**Figure 11.8** Graphical representation of Eqs. (5.17), (11.8a), and (11.12) as functions of $\alpha$ using $\alpha_{\ell 0}$ = − 4.0° and $c_{mc/4}$ = − 0.085 at $\alpha$ = 0°. The plot for Eq. (5.17) is the dashed line, for Eq. (11.8a) the solid line, and for Eq. (11.12) the dash-dot line (magnified by a factor of four). Data points for a NACA-4415 wing section are shown by circles and diamonds. The value of the slope used in both calculated $c_\ell$-curves is 93% of $2\pi$.

bursts, $c_\ell$ may exceed 4.0 as indicated in Figure E11.1b when the boundary layers remain attached. Current research efforts toward building "micro-flying vehicles" that can attain high lift with especially miniaturized flapping units have achieved the robust lift-to-drag ($L/D$) ratios found in nature's flying fauna (see Jones and Platzer 2003).

Analytical solutions to the unsteady small perturbation equation (Liepmann and Roshko 1957) show a "thin hysteresis loop" with only a minor change in the steady-state slope ($c_{\ell\alpha}$), but a loop that extends the stall condition to higher angles of attack. Flapping affects the pressure distribution around the airfoil and thus the instantaneous lift because flapping frequency together with planform configuration govern unsteady effects. Importantly, entrained flow ahead of a flapping wing appears to prevent flow separation, even in the strong adverse pressure gradients encountered at high angles of attack, and this apparently translates into an ability to operate at angles near 45°. Since comparatively long amounts of time are needed for boundary layers to develop, separation can be delayed in airfoils operating under rapidly changing angles of attack.

### 11.7.4 Technology Advances

We have seen that rotating cylinders in ideal flows may develop lift coefficients up to $4\pi$ with the cylinder's chord ($c$) remaining at $2r$ since the radius is not affected by cylinder rotation. For airfoils, however, it is not their chord length but their horizontal projection that defines the lifting planform area, one which changes with $\alpha$ as $c\cos\alpha$. Rotating cylinders, however, were tried and proved impractical for aircraft. Changing the viscous properties in the boundary layers through injection of special gases was another old idea, one that also proved impractical. Much effort has been expended on enhancing lift with engine exit-flow over the wing; this has not been widely adopted because such aircraft proved highly unreliable due to their extra complexity.

Practical schemes for boundary layer control and drag abatement to increase an airfoil's maximum lift performance are in critical demand because of the potential for high payoffs; such research remains ongoing. Modern innovations have accelerated these efforts with recent progress in "artificial intelligence (AI) and deep machine learning." For example, *morphing control surfaces* also known as *load alleviation surfaces* that act as one continuous surface do eliminate drag from seams and hinges. These are *flexible articulating transition surfaces* that unfold on-demand like hand-fans to keep the fixed parts of the wing connected through shape-morphing control surfaces; such accordion-like structures avoid sharp edges as they join flexible flaps to the wing.

Some current research includes:

a) *Morphing Camber:* Wing sections whose camber can be modified by bending a fish-like-skeleton-core without producing wing gaps from rigid control surfaces (i.e. the flap elements in Figure 11.4) when deployed. These may provide more control "authority" for high-speed aircraft especially in very low ***AR*** wings.
b) *Interbonding Polymer Networks*: Shape-changing wing configurations which can, as birds do, alter wing configurations inflight between cruising, high speed sprints, and extreme maneuverability.
c) *Assorted Topics*: New passive methods of promoting turbulence at wing regions under adverse pressure gradients (beyond turbulence trippers or riblets discusses in Chapter 7). Pressure microsensors tied to AI units for precise boundary layer control. Three-dimensional printing of complex aircraft parts, smart structures, and other new manufacturing methods. New analytical work with supercomputers and better wind tunnels with modern instrumentation.

## 11.8 Recapitulation

The topic *of high-lift airfoils* in ideal incompressible flows has been treated separately partly because of its importance during takeoff and landing and, more broadly, for its role on aircraft maneuverability and eventual overall cost. High-lift capabilities affect the ways conventional aircraft can compete with helicopters and other propulsive lift-assisted aircraft for urban trips and on short runways. Present high-lift airfoil designs have much potential for improvement and should benefit from the advent of modern computational tools, new materials, and the latest advances in manufacturing techniques. With all this in mind, we have revisited theoretical performance limits of thin airfoil sections under conditions exceeding angle-of-attack constraints previously considered insurmountable.

Aside from maintaining boundary layers attached, higher angles of attack introduce other technical challenges so examining potential mechanisms that extend stall conditions can be a strong incentive. While the figures in Examples 11.1 and 11.3 represent a purely theoretical behavior for thin airfoils, they do display patterns evident from airfoil data before stall (with the caveat that that different camber shapes affect the magnitude of these numbers as much as the missing boundary layer effects). Discussion of data shown in Figures 11.5 and 11.8 brings out some shortcomings of standard models, and Figure 11.8 clearly exemplifies the defect of using the linear approximation for the trigonometric functions.

We have presented data on airfoil lift vs angle of attack at three places in this chapter: Figure 11.2 corroborates our stated dependance on angle of attack as "$c \cos \alpha$" for that portion of the chord effective in the lift force direction; Figure 11.5 provides two different lessons, one showing that sole deployment of slats extends the linear symmetric range of the airfoil, and the other that experimental results need to be corrected when there is an increase in chord length; Figure 11.8 indicates that the best fit with experimental results is achieved with a lift slope slightly less (93%) than $2\pi$.

Furthermore, all $c_\ell$-graphs in this chapter indicate that Eq. (11.8a) is a truer representation of experimental results than its more common form from thin airfoil theory, Eq, (5.17).

We review here two main deficiencies in the definition of the lift coefficient in Chapter 2. From *dimensional analysis*, we expect that, when a variable is non-dimensionalized using relevant *constants* of the problem, we do obtain more useful parameters; for incompressible flows at low angles of attack $c_\ell = \ell / q_\infty c \approx 2\pi\alpha$ accomplishes that, but only when flaps and slats are fully nested. Moreover, as we discussed in Chapters 5, 9, and in this chapter, this is a very restricted range of operation in aerodynamics. For compressible flows, it is more appropriate to work with the set of dimensionless coefficients for lift, moment, and wave drag given in Table 9.1. For incompressible flows, it is also more appropriate to use the following dimensionless coefficients:

a) To account for increases in chord length (c) with deployed flaps and slats, use equation (11.13)

$$c_\ell = c_{\ell 0}\left(\frac{c_0}{c}\right) \tag{11.13}$$

For $C_L$ an additional correction needed for when these units do not cover entire spans.

b) To account for angles greater than 15° and to arrive at proper ideal-flow parameter limits, we should use Eq. (11.8a) which replaces Eq. (5.17)

$$\boxed{c_\ell = 2\pi\cos\alpha[\sin\alpha - (\cos\alpha)(\tan\alpha_{\ell 0})]} \tag{11.8a}$$

In these relations, $c_{\ell 0}$ and $c_0$ refer to the nested cruise-configuration airfoil values and $\alpha_{\ell 0}$ is the angle of attack for zero lift. For other coefficients related to lift such as the pitching moments, the same modifications apply as given in Eqs. (11.9), (11.10), and (11.12). The coefficient of drag due to friction remains as defined in Chapters 2 and 7. Hypersonic flows and flow conditions inside detached boundary layers introduce additional restrictions that have not been considered.

From a purely theoretical standpoint, it is significant to establish that for thin, symmetrical airfoils in incompressible flow $c_\ell)_{max} = \pi$ at $\alpha_{max} = 45°$ because this value represents a theoretical upper limit. Camber increases the theoretical $c_\ell)_{max}$ to around $2\pi$. Compressibility and thickness do enhance $c_\ell)_{max}$ somewhat. While it has been difficult to keep the boundary-layers attached on common lifting surfaces much beyond $\alpha = 15°$ without implementing sophisticated schemes, in principle, this is a situation that ongoing research efforts should be able to improve. In summary, theoretical upper limits for thin airfoil sectional lift coefficients in ideal incompressible flow, along with the rotating cylinder for comparison, can be given as

| | |
|---|---|
| (a) Rotating circular cylinders | $c_{\ell max} = 4\pi$ |
| (b) Symmetric thin airfoils at $\alpha = 45°$ | $c_{\ell max} = \pi$ |
| (c) Cambered thin airfoils at $\alpha < 20°$ | $c_{\ell max} =$ between $2\pi$ and $2.5\pi$ |

## Problems

**11.1** A two-dimensional symmetric airfoil has a measured $c_{\ell\alpha}$ of 90% theoretical. Calculate $c_\ell$ and $c_{\text{mac}}$ at $\alpha = 18°$.

**11.2** Calculate the total $c_\ell$ for an airfoil with a plain flap of length 0.15c deflect at $\eta = 20°$. The geometric angle of attack is 25°. Assume all boundary layers remain attached.

**11.3** A thin two-dimensional cambered airfoil has a triangular shape. It can be approximated to two straight-line segments as shown in Figure P11.3. Calculate $\alpha_{\ell 0}$ and $c_\ell$ when $\alpha = 6°$.

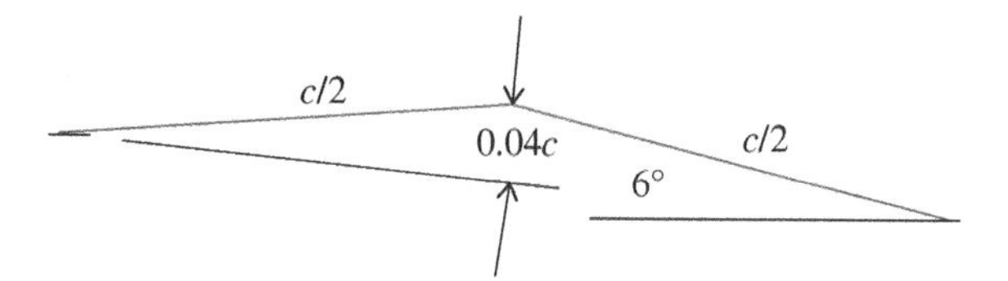


**Figure P11.3**

**11.4** The $c_\ell)_{max}$ for ordinary airfoil sections is about 1.8 but, theoretically, a flat plate can reach the magnitude $c_\ell)_{max} = \pi$, what would be the stall speeds ratio (everything else being equal, same $S$, elevation, weight) of the increase in $c_\ell)_{max}$? Assume that the 2D vs 3D ratios are proportional.

**11.5** Calculate the two-dimensional lift coefficient for a hypothetical cambered airfoil with $\alpha_{\ell 0} = -12°$ operating at $\alpha = 20°$ by both the linear and nonlinear formulations and compare.

**11.6** Calculate the theoretical values of $\alpha_{max}$ and $c_\ell)_{max}$ for the cambered airfoils whose $\alpha_{\ell 0}$ are listed below:

a) $\alpha_{\ell 0} = -4.0°$

b) $\alpha_{\ell 0} = -8.0°$

**11.7** For an aircraft with an elliptical-planform wing of $\boldsymbol{AR} = 10$ calculate the theoretical values of $C_L$ when $\alpha_{\ell 0} = -4.0°$ for $\alpha = 10.0°$ and $20.0°$. Compare with the linear theory of Chapter 6.

## Check Test

**11.1** State three applications for which high angle of attack airfoils are important.

**11.2** Given our stated theoretical upper limits of $c_\ell)_{max}$ ($\pi$ for a symmetrical airfoil and over $2\pi$ for a cambered airfoil), how high is the potential for improving present airfoil performance?

**11.3** A high-lift coefficient can be obtained by increasing the airfoil's angle of attack, deploying flaps/slats, or both. What limits changes to angle of attack, the simplest of the two to implement?

**11.4** Calculate how close Eq. (11.8a) is to Eq. (5.17) (the small angle approximation) when both $\alpha$ and $\alpha_{\ell 0}$ equal 15°.

**11.5** Show that in Eq. (11.8a), $\alpha = \alpha_{\ell 0}$ when $c_\ell = 0$ for any $\alpha$ value (i.e. this result is not limited to small angles).

**11.6** Work problem 11.6(a).

# Appendix A

## Standard Atmosphere SI Units

| Altitude $z$ (m) | Temperature $T$ (K) | Pressure $p$ (N/m$^2$) | Density $\rho$ (kg/m$^3$) | Speed of sound $a$ (m/s) | Kinematic viscosity $\nu$ (m$^2$/s) |
|---|---|---|---|---|---|
| 0 | 288.16 | $1.01325 \times 10^5$ | 1.2250 | 340.29 | $1.4607 \times 10^{-5}$ |
| 600 | 284.26 | $9.4322 \times 10^4$ | 1.1560 | 337.98 | 1.5316 |
| 1 200 | 280.36 | 8.7718 | 1.0900 | 335.66 | 1.6069 |
| 1 800 | 276.46 | 8.1494 | 1.0269 | 333.32 | 1.6869 |
| 2 400 | 272.57 | 7.5634 | $9.6673 \times 10^{-1}$ | 330.96 | 1.7721 |
| 3 000 | 268.67 | 7.0121 | 9.0926 | 328.58 | 1.8628 |
| 3 600 | 264.77 | 6.4939 | 8.5445 | 326.19 | 1.9595 |
| 4 200 | 260.88 | 6.0072 | 8.0222 | 323.78 | 2.0626 |
| 4 800 | 256.98 | 5.5506 | 7.5247 | 321.36 | 2.1721 |
| 5 400 | 253.09 | 5.1226 | 7.0513 | 318.91 | 2.2903 |
| 6 000 | 249.20 | 4.7217 | 6.6011 | 316.45 | 2.4161 |
| 6 600 | 245.30 | 4.3468 | 6.1733 | 313.97 | 2.5509 |
| 7 200 | 241.41 | 3.9963 | 5.7671 | 311.47 | 2.6953 |
| 7 800 | 237.52 | 3.6692 | 5.3818 | 308.95 | 2.8503 |
| 8 400 | 233.63 | 3.3642 | 5.0165 | 306.41 | 3.0167 |
| 9 000 | 229.74 | 3.0800 | 4.6706 | 303.85 | 3.1957 |
| 9 600 | 225.85 | 2.8157 | 4.3433 | 301.27 | 3.3884 |
| 10 200 | 221.97 | 2.5701 | 4.0339 | 298.66 | 3.5961 |
| 10 800 | 218.08 | 2.3422 | 3.7417 | 296.03 | 3.8202 |
| 11 400 | 216.66 | 2.1317 | 3.4277 | 295.07 | 4.1474 |
| 12 000 | 216.66 | 1.9399 | 3.1194 | 295.07 | 4.5574 |
| 12 600 | 216.66 | 1.7654 | 2.8388 | 295.07 | 5.0078 |
| 13 200 | 216.66 | 1.6067 | 2.5835 | 295.07 | 5.5026 |
| 13 800 | 216.66 | 1.4622 | 2.3512 | 295.07 | 6.0462 |
| 14 400 | 216.66 | 1.3308 | 2.1399 | 295.07 | 6.6434 |
| 15 000 | 216.66 | 1.2112 | 1.9475 | 295.07 | 7.2995 |
| 15 600 | 216.66 | 1.1023 | 1.7725 | 295.07 | 8.0202 |
| 16 200 | 216.66 | 1.0033 | 1.6133 | 295.07 | 8.8119 |
| 16 800 | 216.66 | $9.1317 \times 10^3$ | 1.4683 | 295.07 | 9.6816 |
| 17 400 | 216.66 | 8.3115 | 1.3365 | 295.07 | $1.0637 \times 10^{-4}$ |
| 18 000 | 216.66 | 7.5652 | 1.2165 | 295.07 | 1.1686 |
| 18 600 | 216.66 | 6.8859 | 1.1072 | 295.07 | 1.2839 |

*Elements of Aerodynamics: A Concise Introduction to Physical Concepts*, First Edition. Oscar Biblarz.

Companion website: www.wiley.com/go/elementsofaerodynamics

# Appendix B

# Software

Aerodynamics Calculator – Companion to Oscar Biblarz "Elements of Aerodynamics"
https://www.oscarbiblarz.com/aerocalculator

This program reproduces relevant isentropic parameters ratioed to their stagnation value and calculates oblique shocks without the aid of the charts having the normal shock formulations built-in. With the *Aerodynamics Calculator*, the user avoids any need for interpolation or extrapolation and further manipulations to arrive at the desired final answers are easily done with most standard calculators. In addition to being a labor-saving aid, this calculator enables the user to conveniently tackle many problems that require other values of $\gamma$ without defaulting to 1.40.

Sample screen shots:

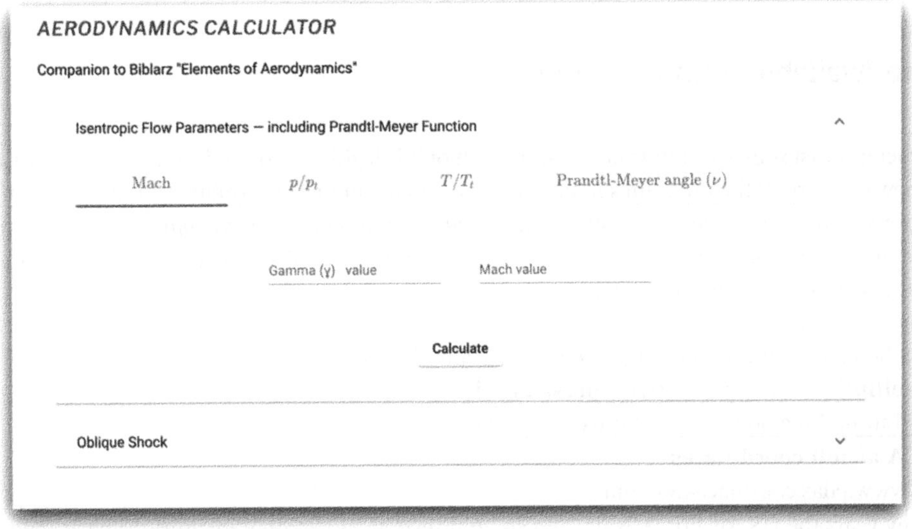

AERODYNAMICS CALCULATOR
Companion to Biblarz "Elements of Aerodynamics"
Isentropic Flow Parameters – including Prandtl-Meyer Function
Mach $p/p_t$ $T/T_t$ Prandtl-Meyer angle ($\nu$)
Gamma ($\gamma$) value Mach value
Calculate
Oblique Shock

This screen shows the isentropic set of parameters. It can accept one of four inputs: Mach number, pressure ratio, temperature ratio, or Prandtl–Meyer angle. The second line shows that the Mach number have been chosen for input and that a value of $\gamma$ is also required. When you click on the "Calculate" button, the program responds by displaying the isentropic results from Eqs. (8.7) to (8.9) and the Prandtl—Meyer angle. Values of $\gamma$ are limited to perfect gases ($1.0 < \gamma < 1.7$).

*Elements of Aerodynamics: A Concise Introduction to Physical Concepts*, First Edition. Oscar Biblarz.

Companion website: www.wiley.com/go/elementsofaerodynamics

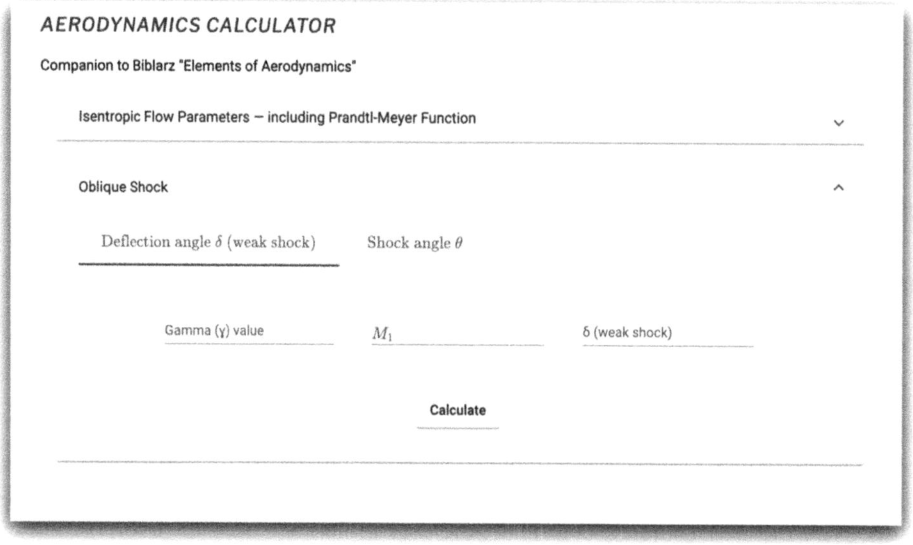

**AERODYNAMICS CALCULATOR**

Companion to Biblarz "Elements of Aerodynamics"

Isentropic Flow Parameters – including Prandtl-Meyer Function

Oblique Shock

Deflection angle $\delta$ (weak shock) | Shock angle $\theta$

Gamma (γ) value | $M_1$ | δ (weak shock)

Calculate

This screen shows the oblique shock screen. It can accept either a deflection angle or a shock angle as inputs, but with the latter it will only output weak-shock results. The second line shows that the deflection angle is the chosen input and that its value and $\gamma$ are required. When you click on the "Calculate" button, the program responds by displaying results from Eqs. (8.21) and (8.22). Values of $\gamma$ are limited to perfect gases ($1.0 < \gamma < 1.7$).

## Software Available from the Internet

The Internet is a vast source of information and much of it is legitimate but it is not always clear to the user how to interpret the program's notation, though many programs available do carry explanatory sections for their use. The listings included here are a small sample of *free software* available at the date of the publication of this book. Neither the author nor the publisher guarantees availability or correctness of any software listed below.

> **Standard Atmosphere**

1976 US Standard Atmosphere http://www.pdas.com/atmos.html

> **Non-elliptic 3D wings in incompressible flow**

Prandtl Lifting Line Solver for 3D unswept wings

> **NACA airfoil coordinates**

https://www.pdas.com/naca456.html

For many years the US Air Force, Navy, NASA, and several educational institutions have sponsored the development of computer software that is useful to aeronautical engineers, airplane designers, and aviation technicians.

> **Public Domain Aeronautical Software (PDAS) Website**

Valuable aeronautical computer programs complete with public domain source code, instructions, and sample cases.

- Table of Contents
- What's New

- Read the Bug List
- Browse the corrected tables for the famous book *Theory of Airfoil Sections*
- Get a copy of all the programs from *Public Domain Computer Programs for the Aeronautical Engineer*
- Read the list of frequently-asked-questions about Public Domain Aeronautical Software (PDAS)
- Download a flyer that describes the programs. You may have the short version (2 pages) or the long version (26 pages)
- Background Information on PDAS (Ralph Carmichael)
- Legal statements about *Public Domain Computer Programs for the Aeronautical Engineer*
- Look at my list of other sources of aeronautical software or aeronautical information or to a page of links to aeronautical information

PDAS was founded to make valuable software available to the aeronautical community for use on desktop computers. These programs include descriptions and complete public domain source code. All are downloadable free of charge from this website. Many programs have sample cases (both input and output). The source code is not copyrighted and may be used in whole or in part in any of your aeronautical studies. Most were developed under NASA or DOD sponsorship, but some are contributions from individual authors and all have significant value added by PDAS.

> **NASA Glenn Research Center (GRC) Website**

https://www.grc.nasa.gov/www/k-12/airplane/short.html

In an effort to foster hands on, inquiry-based learning in science and math, the NASA Glenn Research Center has developed a series of interactive computer programs for k-12 students. All of the programs are Java applets which run in a browser, online, over the World Wide Web. The programs can also be downloaded to the computer for use without being online. The programs are in the public domain and are constantly being modified and upgraded based on users input. Glenn Research Center has also developed a series of Beginner's Guides that accompany each of the software packages to explain the science and math.

# Appendix C

# Equations for Chapters 5 and 6

## The Fourier Sine-Series Formulation

In order to formulate vortex distributions in cambered airfoils and in non-elliptic wings, a *Fourier series representation* is particularly useful. This is because, besides using well-known techniques, this description needs the least number of terms for cases of interest.

Here we require a change of variable for both the *chord* and the *span* coordinates from $x$ or $y$ to the polar coordinates $r$ and $\theta$.

$$x = \frac{1}{2}c(1 - \cos\theta) \text{ and } y = \frac{1}{2}b(1 - \cos\theta) \tag{C.1}$$

Note that the angle $\theta$ is hinged at mid-chord for Part A (below) or mid-span for Part B (below) so that the limits $(0, c)$ or $(-b/2, b/2)$ both become $(0, \pi)$. One advantage of a Fourier series approach is evident in the fast convergence of its resulting terms.

A) For a cambered airfoil of given sectional slope $\left.\dfrac{dz}{dx}\right)_w$, the following formulations apply:

$$\gamma(\theta) = 2V_\infty \left[A_0 \frac{1 + \cos\theta}{\sin\theta} + \sum_{n=1}^{\infty} A_n \sin n\theta\right] \tag{C.2}$$

$$\text{Kutta} - \text{condition } \gamma(\pi) = 0 \tag{C.3}$$

$$\left.\frac{dz}{dx}\right)_w = (\alpha - A_0) + A_1 \cos\theta + A_2 \cos 2\theta + \ldots \tag{C.4}$$

The coefficients $A_0$, $A_1$, and $A_2$ represent a $\gamma(\theta)$ distribution that satisfies the condition of having the flow parallel to the camber line. In Chapters 5, we introduced the resulting equations without using $A$ coefficients since only the first three are needed. Here we write down their equivalent forms.

$$A_0 = \alpha - \frac{1}{\pi}\int_0^\pi \left(\frac{dz}{dx}\right)_w d\theta \tag{C.5}$$

$$A_n = \frac{2}{\pi}\int_0^\pi \left(\frac{dz}{dx}\right)_w \cos(n\theta) d\theta \tag{C.6}$$

*Elements of Aerodynamics: A Concise Introduction to Physical Concepts*, First Edition. Oscar Biblarz.

Companion website: www.wiley.com/go/elementsofaerodynamics

$$c_\ell = 2\pi A_0 + \pi A_1 \text{ and } c_{m0} = -\frac{1}{2}\pi\left(A_0 + A_1 - \frac{1}{2}A_2\right) \tag{C.7}$$

B) Finite wings with an arbitrary circulation distribution

An arbitrary $\Gamma(\theta)$-distribution may be represented as shown in equation (C.8), where $\theta$ is the polar coordinate with its corresponding $r$ pivoting from $y = 0$. As stated, the angle $\theta$ sweeps the span counterclockwise making $\theta = 0$ at $b/2$ and $\pi$ at $-b/2$.

$$\Gamma(\theta) = 2bV_\infty \sum_{n=1}^{N} A_n \sin(n\theta) \tag{C.8}$$

With this change of variable, the first term in the series is the elliptical distribution, and it can be shown that only odd terms in the Fourier series survive under symmetrical lift loadings:

$$\Gamma(\theta) = 2bV_\infty [A_1 \sin\theta + A_3 \sin(3\theta) + A_5 \sin(5\theta) + A_7 \sin(7\theta) + \ldots] \tag{C.9}$$

$$C_L = \pi A_1 \boldsymbol{AR} \tag{C.10}$$

$$C_{Di} = \pi\,\boldsymbol{AR} \sum_{n=1}^{N} nA_n^2 \tag{C.11}$$

Properly designed trapezoidal wings add less than 10% to the $C_{Di}$ of elliptical wings. Solutions to these equations require additional steps and a matrix inversion (see Anderson 2017; Bertin and Cummings 2013; Kuethe and Chow 1998; McCormick 1979; Schlichting and Truckenbrodt 1979; Shevell 1983).

For an untwisted rectangular wing of given ***AR*** at an angle of attack $\alpha$, the $A$s are found by solving equation (C.12). Lack of twist means the angle of attack is same everywhere as $\alpha$ and we use $a_0 = 2\pi$. At every station, $\theta = \cos^{-1}\left(\frac{2y}{b}\right)$. The equation that must be satisfied is shown in Kuethe and Chow (1998) to be

$$\sum_{n=1}^{\infty} A_n \sin n\theta \left(1 + \frac{n\pi}{2\boldsymbol{AR}\sin\theta}\right) = \alpha \tag{C.12}$$

Only odd $n$-terms apply when there is loading symmetry, and four of them along the half-span turn out to be generally sufficient. Numerical results from an example in Kuethe and Chow (1998) show that $C_{L\alpha}$ can be 4% lower than the elliptic value and that $C_{Di}$ can be 5% higher than in the elliptic span loading. These results confirm that nearly elliptic ideal results can be obtained with rectangular configurations.

- Prandtl Lifting Line Solver for 3D unswept wings (MS Windows software, downloadable)

# Selected References

## Aerodynamics

Abbott, I.H. and von Doenhoff, A.E. (1949). *Theory of Wing Sections*. New York: Dover Publications.
Anderson, J.D. (1978). *Introduction to Flight*. New York: McGraw Hill.
Anderson, J.D. (2017). *Fundamentals of Aerodynamics*, 6e. New York: McGraw Hill.
Bertin, J.J. and Cummings, R.M. (2013). *Aerodynamics for Engineers*, 6e. Prentice Hall.
Glauert, H. (1947). *The Elements of Aerofoil and Airscrew Theory*. Cambridge: Cambridge University Press.
Hoerner, S.F. (1975). *Practical Information on Fluid Dynamic Lift*. Self-published.
Houghton, E.L. and Carpenter, P.W. (1993). *Aerodynamics for Engineering Students, 4e*. New York: Halsted Press.
Katz, J. and Plotkin, A. (2000). *Low Speed Aerodynamics*, 2e. Cambridge: Cambridge University Press.
Kuethe, A.M. and Chow, C.Y. (1998). *Foundations of Aerodynamics*, 5e. John Wiley & Sons.
McCormick, B.W. (1979). *Aerodynamics, Aeronautics, and Flight Mechanics*. New York: John Wiley & Sons.
Prandtl, L. and Tietjens, O.G. (1934a). *Applied Hydro- and Aeromechanics*. New York: Dover Publication.
Prandtl, L. and Tietjens, O.G. (1934b). *Fundamentals of Hydro- and Aeromechanics*. New York: Dover Publication.
Schlichting, H. and Truckenbrodt, E. (1979). *Aerodynamics of the Airplane*. New York: McGraw Hill.
Shevell, R.S. (1983). *Fundamentals of Flight*. Englewood Cliffs, NJ: Prentice-Hall.
Tennekes, H. (1997). *The Simple Science of Flight*. Cambridge: The MIT Press.

## Fluid Mechanics

Landau, L.D. and Lifshitz, E.M. (1959). *Fluid Mechanics*. Oxford: Pergamon Press.
Munson, Y. (2021). *Okiishi's Fundamentals of Fluid Mechanics*, 9e. Hoboken, NJ: John Wiley & Sons.
Streeter, V.I. and Wylie, E.B. (1985). *Fluid Mechanics*, 8e. New York: McGraw-Hill.

## Gas Dynamics

Liepmann, H.W. and Roshko, A. (1957). *Elements of Gas Dynamics*. New York: John Wiley & Sons.
Saad, M.A. (1985). *Compressible Fluid Flow*. Englewood Cliffs, NJ: Prentice-Hall.
Shapiro, A.H. (1953). *The Dynamics and Thermodynamics of Compressible Fluid Flow*. New York: John Wiley & Sons.

*Elements of Aerodynamics: A Concise Introduction to Physical Concepts*, First Edition. Oscar Biblarz.

Companion website: www.wiley.com/go/elementsofaerodynamics

Zucker, R.D. and Biblarz, O. (2020). *Fundamentals of Gas Dynamics*, 3e. Hoboken, NJ: John Wiley & Sons.
Zuckrow, M.J. and Hoffman, J.D. (1976). *Gas Dynamics*, vol. I. New York: John Wiley & Sons.

## Hypersonics

Anderson, J.D. (1989). *Hypersonic and High Temperature Gas Dynamics*. New York: McGraw Hill.

## Propulsion

Archer, R.D. and Saarlas, M. (1996). *An Introduction to Aerospace Propulsion*. Upper Saddle River, NJ: Prentice Hall.
Hill, P.G. and Peterson, C.R. (1962). *Mechanics and Thermodynamics of Propulsion*, 2e. Reading, MA: Addison-Wesley.
Sutton, G.P. and Biblarz, O. (2017). *Rocket Propulsion Elements*, 9e. New York: John Wiley & Sons.

## Thermodynamics

Dittman, R.H. and Zemansky, M.W. (1996). *Heat and Thermodynamics*. New York: McGraw-Hill.
Moran, M.J. and Shapiro, H.N. (1999). *Fundamentals of Engineering Thermodynamics*. New York: John Wiley & Sons.
Sonntag, R.E., Borgnnake, C., and Van Wylen, C.J. (1997). *Fundamentals of Thermodynamics*, 5e. New York: John Wiley & Sons.

## Transonics

Guderley, K.G. (1962). *The Theory of Transonic Flow*. London: Pergamon Press.
Ramm, H.J. (1990). *Fluid Dynamics for the Study of Transonic Flow*. Oxford: Oxford University Press.

## Other Publications

Biblarz, O. (1976). *Phase Plane Analysis of Transonic Flows*, AIAA-76-332,.
Biblarz, O. and Priyono, E. (1994). *Transonic Pressure Drag Coefficient for Axisymmetric Bodies*, ICAS-95-2.5.2.
Jones, K. D. and Platzer, M. F. (2003). *Experimental Investigation of the Aerodynamic Characteristics of Flapping-Wings Micro Air Vehicles*, AIAA-2003-0418.

# Answers to Selected Problems

## Chapter 1

**1.2** −10.33 m

**1.3** $R = 296.72$ N-m/kg-K, $\gamma = 1.40$

**1.4** Density ratio = 5.66, pressure ratio = 11.31

**1.5** Isothermal atmosphere = 0.268 bar, linear decrease = 0.222 bar

## Chapter 2

**2.2** $V_\infty$, $\mu_\infty$, and $c$

**2.3** $a_\infty = \mathrm{f}(\frac{p_\infty}{\rho_\infty}, \gamma)$

**2.4** $V_{\infty mars} = 4.28 V_{\infty earth}$

## Chapter 3

**3.1** $\emptyset = (x^2 - y^2) + xy + \text{constant}$, $\psi = \frac{1}{2}(y^2 - x^2) + 2xy + \text{constant}$

**3.3** Irrotational and incompressible

**3.4** $\emptyset = U_0 x + V_0 y + \text{constant}$, $\psi = U_0 y - V_0 x + \text{constant}$

**3.5** $\psi = xy + \text{constant}$

*Elements of Aerodynamics: A Concise Introduction to Physical Concepts*, First Edition. Oscar Biblarz.

Companion website: www.wiley.com/go/elementsofaerodynamics

## Chapter 4

**4.1** $\rho g\Delta z = 13\,608\ \text{N/m}^2$, $(p_1 - p_2) = 13\,607\ \text{N/m}^2$

**4.5** $V_1 = 84.22$ m/s, maximum

**4.7** $p = 40.69$ psi, $p_t = 54.11$ psi

**4.8** $p_{t1} = 9.378 \times 10^4\ \text{N/m}^2$, $p_{t2} = 1.035 \times 10^5\ \text{N/m}^2$

## Chapter 5

**5.1** $c_\ell = 0.746$, negative

**5.3** $c_\ell = 2.375$

**5.4** $c_\ell = 1.48$

**5.6** $\Delta c_\ell = 0.320$, $\Delta c_{mac} = -0.08$

**5.7** $c_\ell = 1.52$, $c_{mc/4} = -0.165$

**5.8** $c_\ell = 1.8$

## Chapter 6

**6.1** $C_{L\alpha} = 4.93$, $C_{Di} = 0.0326$

**6.2** $\alpha = 7.85°$

**6.4** (a) $S = 9.50\ \text{m}^2$ (b) $S = 18.3\ \text{m}^2$

**6.5** $C_D = 0.0561$

## Chapter 7

**7.1** (i) $D = 8.953$ N (ii) $D = 62.24$ N

**7.2** $c = 53.40$ cm

**7.5** The lift force measured in the air tunnel will be 1/4 of that in the water tunnel, while the air velocity would have to be almost 15 times faster. To calculate more explicit values, we would need to specify the airfoil dimensions.

**7.6** $C_D = 4.70 \times 10^{-2}$

## Chapter 8

**8.1** $Ma = 0.586$, $p/p_t = 0.7925$

**8.2** $C_{pt} = 1.089$, $C_{p,\,cr} = -1.69$

**8.3** $\ell = 49.50$ kN/m, $d = -12.62$ kN/m

**8.4** (c) $\ell = 94$ kN/m, $d = 109$ kN/m

**8.5** $d_2 = -2.2$ kN, $d_4 = +0.53$ kN

## Chapter 9

**9.2** $Ma_\infty = 0.2$ $c_\ell = 0.671$ and $Ma_\infty = 0.7$ $c_\ell = 0.921$

**9.3** Ackeret: top $p/p_\infty = 1.16$ and bottom $p/p_\infty = 1.48$. Oblique shock: top $p/p_\infty = 1.17$ and bottom $p/p_\infty = 1.58$

**9.4** $Ma_\infty \geq 1.95$

**9.5** $T = 1517$ lbf

**9.7** (a) fraction = 90% (b) $c_{\ell\alpha} = 1.03$

**9.8** Small perturbation $C_p = -0.0742$ Prandtl–Meyer $C_p = -0.0673$

**9.9** $C_L = 1.10$

**9.10** $\alpha_2 = 4.48°$

**9.11** $C_L = 0.748$

## Chapter 10

**10.1** At $Ma_1 = 1.1$ it is close but diverges rapidly as the incoming Mach number increases.

**10.3** $V_\infty = 1814.7$ m/s

**10.4** $p_3/p_1 - p_2/p_1 = 2.043 - 0.444 = 1.60$

**10.5** Oblique shock $c_\ell = 0.535$ Newtonian $c_\ell = 0.433$

## Chapter 11

**11.1** $c_\ell = 1.66$ $c_{mac} = 0$

**11.2** $c_\ell = 3.51$

**11.3** $c_\ell = 0.970$ $\alpha_{\ell 0} = -2.92°$

**11.4** About a 25% decrease in stall speed

**11.5** The difference is 9.4% with Eq. (11.8a) having a lower value than Eq. (5.17)

**11.6** $\alpha_{max} = 41°$ and $c_{\ell max} = 3.61$

**11.7** For $\alpha = 10.0°$, $C_{L\alpha} = 4.86$ $C_L = 1.16$. For $\alpha = 20.0°$, $C_{L\alpha} = 3.96$ $C_L = 1.52$
Linear: $C_{L\alpha} = 5.24$, $C_L = 1.28$ at 10° and 2.19 at 20°

# Index



*Elements of Aerodynamics: A Concise Introduction to Physical Concepts*, First Edition. Oscar Biblarz.

Companion website: www.wiley.com/go/elementsofaerodynamics

Printed and bound by CPI Group (UK) Ltd, Croydon, CR0 4YY
16/04/2025
14658590-0001